GLOBAL ENVIRONMENTAL POLICIES

GLOBAL ENVIRONMENTAL POLICIES

Ravinder Dhanai

CYBER TECH PUBLICATIONS
4264/3, Ansari Road, Daryaganj, New Delhi-110002 (India)
Ph.: 011-23244078, 011-43559448 Fax: 011-23280028
E-Mail: cyberpublicationsdelhi@yahoo.com
Website: www.cybertechpublications.com

GLOBAL ENVIRONMENTAL POLICIES

Ravinder Dhanai

First Edition 2010

Published by :

G.S. Rawat for **Cyber Tech Publications**
4264/3, Ansari Road, Daryaganj, New Delhi-110002 (India)
Ph.: 011-23244078, 011-43559448 Fax: 011-23280028
E-Mail: cyberpublicationsdelhi@yahoo.com
Website: www.cybertechpublications.com

Preface

This book covers a broad spectrum of different types of countries, ranging from advanced industrial countries, newly industrializing and so-called transition countries to developing countries. While highly developed countries and the most advanced frontrunners in environmental policy were, not unintentionally, over-represente stable environmental policy.

This means that, in principle, it is necessary to include the implementation level or even full policy cycle in the analysis of national environmental policies if the quality of environmental capacities is to be correctly assessed. The case studies presented here and in volume 1 may also be considered a first necessary step towards the design of such a complex and elaborate integrative analytical approach. In developing and transition countries, for example, insufficient monitoring and reporting capacities as well as underdeveloped democratic structures and processes often play a much more important role than in advanced countries. The same is true of capacity deficits at the implementation level. Such deficits may lead to insubstantial real outcomes with respect to emission reduction or environmental quality in spite of sufficient institutional capacities at the national level.

This not only offers a broader empirical basis for generalizing statements, but also the opportunity to make problems and capacity deficits visible that often play only a secondary role in advanced industrialized countries. In particular, it permits identification of the kinds of capacity that need particular strengthening if an effective global environmental policy is to be attainable. This information is particularly relevant for international organizations, whose strategies and support often do not sufficiently reflect such factors.

—Author

CONTENTS

1

AN INTRODUCTION

It is evident that the growth of law and institutions keeps pace with the changing needs of international society. The need for global regulatory efforts has been mooted, in addition to efforts at the national level, to address envi- ronmental problems with global dimensions.

This has, in turn, put to test the traditional rules and principles of international law. It is now well recognised that the legal underpinnings of the current drive for the legal protection of the global environment remain embedded in the general principles of international law.

In recent years, how- ever, the increasing reliance upon Multilateral Environmental Agreements (MEAs) as a pivotal mode of international environmental law-making has long-term implications for the process of centralised legalisation on environmental issues as well as for the body of international law itself.

Since the establishment of the United Nations, inter- national institutions have provided useful forums for international cooperation on a variety of issues. The growth of functional international organisations (specialised agencies), functional commissions and regional commissions as well other programmes within the UN system provide living testimony to the craving for institutionalisation of international cooperation. The web of institutionalisation is also growing in the field of international environmental law. In fact, the recent World Summit on Sustainable Development (WSSD) was expected to give attention to the debate on strengthening the institutional architecture for international environmental governance (IEG). The debate in this respect, however, is not confined to the UN's environment programme (UNEP). It has drawn the attention of States, intergovernmental organisations and civil society.

As a plenary organ of the United Nations, the General Assembly has provided crucial guidance to the whole process even as the technique of 'global conferencing' has become more popular. In the environmental arena, the General Assembly has acted as 'conductor of a grand orchestra' that provides political guidance to States, notwith- standing the inbuilt Charter limitation that it can only make 'recommendations' through its resolutions.

In the environmental arena, the Assembly provides guidance to its member States through its subsidiary organ, UNEP. This entity has acted as a catalyst for the environmental lawmaking process. This again presents an anomalous situation because for such initiatives, another subsidiary organ of the Assembly – International Law Commission – has been sidestepped. The Assembly has adopted, often without a vote, resolutions comprising constituent instruments for specialised environmental institutions such as UNEP and the Commission on Sustainable Development (CSD). In this context, the focus of this article is on UNEP's birth, mandate, contribution in the development of inter- national environmental law, its post-Rio political neglect, intergovernmental efforts at 'revitalisation' as well as its future direction in the context of the current debate on 'international environmental governance' (IEG), among others, within the UN system, governments and civil society.

PREMIER ENVIRONMENTAL ENTITY

UNEP was established by the UN General Assembly in 1972 as a follow-up to the first UN Conference on Human Environment (UNCHE) to meet the 'urgent need for a permanent institutional arrangement within the United Nations system for the protection and improvement of the environment.'

The General Assembly gave a mandate to the UNCHE to lay down guidelines for action, by national governments as well as international organisations, on broad topics of significance. Being the first major UN ef- fort to diagnose global environmental problems, the UNCHE was expected to address both legal and institutional responses to the environmental challenge. Thus the Stockholm Conference may well be regarded as a major milestone and catalyst in heralding the process of 'greening' of international law. UNEP's constituent instrument per se does not mention its role in international environmental law-making. This, it seems, has now come to be regarded as its 'cornerstone'.

The innovative exercise of such implied power gives latitude to the institution as well as facilitates in actual governance by giving effect to such a mandate. Most institutions come across a situation during their lifespan when they are required to exercise certain powers or functions so as to 'effectuate the general purposes for which (they have been) created, though (their) constitutive instrument does not expressly provide for such a specific power or function'.

State practice shows that, for maximum clarity, States do try to put things in the written form. However, many of the things that are not clarified in the constituent instru- ment can be presumed, if they are essential for its raison d'être. Therefore, the theory of 'implied power' has come to be widely accepted. This was affirmed by the International Court of Justice (ICJ) in the Reparations case in this much-quoted passage:

Under international law, the Organisation must be deemed to have those powers which, though not expressly provided in the Charter, are conferred upon it by necessary implication as being essential to the performance of its duties. This [is a] principle of law. An exercise of such a power may depend upon several factors, such as the intention of the parties involved, the scope of the law's objectives and purposes, the ambit of its expressed powers, the historical background and the need to adapt to changing times. In the specific case of UNEP, the attitude of States, especially in the Governing Council, can act as a barometer of endorsement for the exercise of such implied powers. In this sense, as creatures of institutionalised cooperation, international environmental institutions (IEIs) evolve as dictated by the interests and expectations of the sovereign States.

Thus the successful carving out of such a role by UNEP is a manifestation of innovative exercise of implied power. The drafters of UNEP's constituent instrument (General Assembly resolution 2997/ XXVII) might have envisaged such a role but left it as amorphous, probably so as not to cause panic among other high-profile 'specialised agencies' (some of which were already dealing with environmental issues). It seems to have been tacitly understood and accepted, at the time of adoption of this resolution, that UNEP would undertake such a role. Such an implicit role for UNEP could have been necessary, especially due to moves to deny status of a specialised agency to UNEP, efforts at the time to establish an environmental agency outside the UN system, as well as concerns and suspicions of a large number of developing countries regarding practical implications of conceding such a role to a UN programme. It seems that the subsequent attitude and State practice has come to accept the role that UNEP has performed under its Montevideo Programme in view of an absence of any other institutional actor that can undertake such a role.

'CATALYST' FOR LAW-MAKING

An important characteristic of IEIs is that they stem from international instruments (not necessarily creating 'legal' obligations for the States). Once formed, they con- tribute to the law-making process itself. This process happens in a subtle way. In the absence of a supra-legislature at the international level, States inevitably come to rely upon an IEI to provide an institutionalised platform and, in turn, facilitate international cooperation. Such international institutions, especially in the environmental field, have come to play the role of a catalyst in the law-making process. As one of the former Executive Directors of UNEP observed:

> To be able to catalyse action, an organisation must have full command of relevant information, the capacity to interpret data in the development of knowledge and the ability to translate that knowledge and wisdom which informs policy development. These, together, form the intellectual basis of an 'organisation's authority' for management of environmental disputes.

The precise nature of the role played by an IEI will depend upon the mandate of its constituent instrument, the exigencies of situation and the expectation of sovereign States from it. Interestingly, if States are willing to place their confidence in it, the lack of an explicit mandate for the purpose may not per se remain a handicap for an IEI. In UNEP's case, it has exercised roles that were both its own as well as in conjunction with other intergovernmental institutions. The launching of an ambitious Montevideo Programme has served as an important part of UNEP's initiatives for a coherent programmatic approach to law-making on sectoral environmental issues.

The Governing Council of UNEP, whose decisions have also provided an impetus to the States to carry out intergovernmental negotiations, has duly supervised this process. That is also scrutinised by the General Assembly while considering the annual reports of the Governing Council. It has resulted in a fine collection of soft law instruments containing principles, rules, guidelines and recommendations, as well as initiating and contributing to negotiations and the adoption of several MEAs. Such a task performed by an IEI such as UNEP underscores its vital role in enhancing international environmental cooperation as well as acting as a catalytic agent for interna- tional environmental law-making.

The Cooperative Forestry Assistance Act (CFAA) of 1978, as amended, (16 U.S.C. 2101 et. seq.) recognizes that the majority of the Nation's productive forest lands are in private ownership; that private landowners are facing increased pressure to convert their forest lands to other uses; that greater population density, user demands and restrictions on Federal and other public lands are placing increased pressures on private lands to provide a wide variety of products and services from working forests including timber and other forest commodities, fish and wildlife habitat, watershed function and water supply, aesthetic qualities, historical and cultural resources, and recreational opportunities; and that good stewardship of privately held forest lands requires a long-term commitment that can be fostered through a partnership of Federal, State, local government and individual efforts. In 1990, the Forest Legacy Program (FLP) was established to promote the long-term integrity of forestlands. The Secretary was directed to establish the FLP in cooperation with State, regional, and other units of government. In carrying out this mandate, the Secretary of Agriculture is authorized to acquire lands and interests in lands in perpetuity for inclusion in the FLP. Landowner participation in the FLP, including the sale of lands and interests in lands, is entirely voluntary. The Program is implemented through State participation, consistent with these National FLP guidelines, and as described in each State Assessment of Need.

The FLP goals and objectives are accomplished through Forest Service (FS) cooperation with State partners, Federal agencies, local units of government, forest landowners and other partners. The FLP identifies and protects environmentally important private forestlands that are threatened by conversion to nonforest uses and provides the opportunity for continuation of traditional forest uses, such forest management activities and outdoor recreation.

The guidelines are organized in three parts:

PART 1-General Program Guidelines: Program direction applicable to all aspects of the FLP.

PART 2-State Grant Program Guidelines: Program direction applicable to States and Forest Service (FS) Regions/Area/IITF where a State has elected the State grant option and where ownership of lands or interests in lands is vested in a State or subdivision of a State.

PART 3-Federal Acquisition Program Guidelines: Program direction applicable to States and FS Regions/Area/IITF selecting the Federal acquisition and ownership process, where ownership of lands or interests in lands is vested in the United States (U.S.).

AUTHORITY AND PURPOSE OF THE FOREST LEGACY PROGRAM (FLP)

A. Authority

The Cooperative Forestry Assistance Act (CFAA) of 1978, as amended, (16 U.S.C. 2101 et. seq.) provides authority for the U.S. Secretary of Agriculture (Secretary) to provide financial, technical, educational, and related assistance to States, communities, and private forest landowners. Section 1217 of Title XII of the Food, Agriculture, Conservation and Trade Act of 1990 (P.L. 101-624:104 stat.3359; 16 U.S.C. 2103c), also referred to as the 1990 Farm Bill, amended the CFAA and directs the Secretary to establish the FLP to protect environmentally important forest areas that are threatened by conversion to nonforest uses.

This authority continues indefinitely. Through the 1996 Farm Bill (Federal Agricultural Improvement and Reform Act of 1996; P.L. 104-127; Title III-Conservation; Subtitle G-Forestry; Section 374, Optional State Grants for Forest Legacy Program), the Secretary is authorized, at the request of a participating State, to make a grant to the State to carry out the FLP in that State, including the acquisition by the State of lands and interests in lands.

B. Purpose of the Forest Legacy Program

The purpose of the FLP is to ascertain and protect environmentally important forest areas that are threatened by conversion to nonforest uses.

FLP seeks to promote forestland protection and other conservation opportunities. Such purposes shall include the protection of important scenic, cultural, fish, wildlife and recreational resources, riparian areas and other ecological values. Traditional forest uses, including timber management, as well as hunting, fishing, hiking, and similar recreational uses are consistent with purposes of the FLP. Both purchased and donated lands and interests in lands through the use of conservation easements and fee-simple purchase are used to acquire forested land meeting Forest Legacy purposes from willing sellers or donors.

C. Delegations of Authority

The Secretary has delegated authority to administer all aspects of the FLP to the Under Secretary for Natural Resources and Environment (7 CFR 2.20(a)(2)(xvi)) who in turn has delegated the authority to the Chief of the Forest Service (7 CFR 2.60(a)(16). Delegations only apply within the U.S. Department of Agriculture and its agencies. The role of State and Regional programs, and the right of States to elect the State Grant Option, are contained in the authorizing statute and these program implementation guidelines.

DESCRIPTION OF TERMS AND ABBREVIATIONS

Assessment of Need (AON) is a document produced by a State, or a federally recognized Indian Tribe, in consultation with the State Forest Stewardship Coordinating Committee (SFSCC). The AON contains the an assessment of the forests and forest uses, a description of forces that are converting forests to nonforest uses, describes Eligibility Criteria developed by the State to identify important forest areas to be proposed as Forest Legacy Areas (FLA), and acts as a guide to implementation of FLP in the State.

Assessment of Need (AON) Amendment is a document produced by a State to amend their AON, to add or delete Forest Legacy Areas (FLA), or to modify the Eligibility Criteria. CFAA is the Cooperative Forestry Assistance Act of 1978, P.L. 95-313, 92 stat. 365, 16 U.S.C., 2101 et seq. (as amended through P.L. 107-195, June 16, 2002).

Conservation Easement is a legal agreement a property owner makes with a governmental entity or a nonprofit organization to restrict activities allowed on the land in order to protect specified conservation values. Easement restrictions are tailored to the particular property and to the interests of the individual landowner. All FLP conservation easements are held in perpetuity.

Eligibility Criteria are a set of factors developed by the State lead agency, in consultation with the State Forest Stewardship Coordinating Committee (SFSCC), to evaluate geographic areas to determine if they contain significant environmental values to be considered an 'important forest area' and contain "threats" of conversion to be eligible as a Forest Legacy Area (FLA).

Federal Appraisal Standards are those standards contained in the publication entitled "Uniform Appraisal Standards for Federal Land Acquisitions: Interagency Land Acquisition Conference, 2000," or subsequent amendments or updates.

Forest Legacy Area (FLA) is a geographic area with important forest and environmental values, that satisfies identified Eligibility

Criteria and has been delineated, described, and mapped in a State's AON for the FLP. Acquisition of lands and interests in lands for the FLP can only occur within approved FLAs.

Forest Legacy Area (FLA) Boundary Adjustment is a minor change to an existing FLA to create a more logical or manageable boundary.

Forest Legacy Program (FLP) Project is an individual or series of land or interest in land acquisition transaction(s). The transaction(s) can be on an individual tract or multiple tracts in a distinct geographical area. A FLP project relates to a single funding event in a given fiscal year. FLP projects can have a single parcel that can be completed at one closing or more than one parcel that can be completed in a succession of closings. If a successive FLP project is proposed on a parcel or in a distinct geographic area each transaction is treated as an independent unit in the project selection process and funding is not guaranteed.

Forest Service (FS) is the United States Department of Agriculture Forest Service. Forest Service Region/Area/IITF refers to the field units of the Forest Service responsible for FLP management and oversight within the Forest Service Regions, Northeastern Area (Area) or International Institute of Tropical Forestry (IITF).

Forest Stewardship Plans, or multi-resource management plans, are prepared with the purpose of achieving long-term stewardship of forestland. Such plans identify landowner objectives and describe actions to protect and manage soil, water, range, aesthetic quality, recreation, timber, and fish and wildlife resources, and other conservation values identified on the tract. Plans are to be prepared by a professional resource manager. A Forest Stewardship Plan that meets the requirements of the Forest Stewardship Program or a multi-resource management plan is required for FLP qualification. The State Forester or equivalent, or their designee must approve the plan.

Full Fee Purchase is a land conveyance where a purchaser acquires all rights, title and interest in a property from a seller or owner. It is also known as fee simple or fee acquisition. Geographic Regions are the collection of States that makeup the National Association of State Foresters (NASF) Regions. The three regions are: North (consisting of the States within the FS Northeastern Area), South (consisting of all the States within the FS Southern Region, and the Territories of the International Institute of Tropical Forestry), and the West (consisting of all the States within the FS Northern, Rocky Mountain, Intermountain, Southwestern, Pacific Southwest (including the Commonwealth of the Northern Mariana Islands, Guam and American Samoa), Pacific Northwest and Alaska Regions.

Indirect costs relate to costs of the management and administration of the FLP. Indirect costs, unlike salary, which is a direct cost, are defined as costs not readily assignable to a specific legacy acquisition. In-kind contributions are non-cash contributions, including third-party contributions. In-kind contributions must be expenses necessary to accomplish program activities, and allowable if the Federal Government were required to pay for them. Interests in Land are a right, claim, or legal share in real property that are less than the full title. Land Trust is a nonprofit organization, as described in 501(c) of the Internal Revenue Code of 1986, that protects land by working with landowners who wish to donate or sell fee title or conservation easements to maintain conservation values associated with the land.

Market Value is the amount in cash, or in terms reasonably equivalent to cash, for which in all probability the property would be sold by a knowledgeable owner willing but not obligated to sell to a knowledgeable purchaser who desires but is not obligated to buy. (Uniform Appraisal Standards for Federal Land Acquisitions: Interagency Land Acquisition Conference, 2000, p.4.) Multi-State Entity is a government-established organization involving two or more States or Indian tribes whose jurisdiction encompasses all or portions of the land area of an FLA(s). National Association of State Foresters (NASF) is the organization representing State forestry organizations in all 50 States, the territories and the District of Columbia. Nonfederal Cost Share refers the nonfederal cost-share required to receive FLP funding. There are three main categories of activities that meet this requirement:

(1) the value of land, or interests in land, dedicated to the FLP that is not paid for by the Federal government,

(2) nonfederal costs associated with program implementation,and

(3) other nonfederal costs associated with a grant or other agreement which meets FLP purposes. The nonfederal cost-share must be documented, and in the case of a grant, must meet the timing, terms and conditions of the grant.

Nonforest Uses

Noncompatible-nonforest uses are uses of the land inconsistent with maintaining forest cover including, but not limited to, activities that result in extensive surface disturbance such as residential subdivisions, commercial development, and mining. These uses generally should be excluded from FLP conservation easements or land purchases. FLP funds should only be used on parcels with forestland as defined in the State's AON. Compatible-nonforest uses are nonforest uses of the land that may be compatible with forest uses as part of an undeveloped landscape, including cultivated farmland, pasture, grassland, shrubland, open water, and wetlands. These nonforest uses should be less than 25 percent of the total area. Forest Legacy funds should only be used on parcels with forestland as defined in a State's AON. Other funding sources may be used to protect nonforested areas on those parcels with less than the minimum required forest cover.

Nontrust Allotment Lands are privately owned fee simple lands owned by tribal members and if they are forested, are eligible for the FLP when they are located within an approved FLA. Trust lands and reservations are already protected through the trust relationship between the U.S. Department of the Interior and the tribe and are ineligible for the FLP. Pass-through describes a land transaction whereby a third party, such as a land trust, acquires interests in lands with the intent to convey such interests to a unit of government. The transaction can include a full or partial donation, or sale at market value. Payment in Lieu of Taxes (PILT) is made by tax-exempt entities, including the Federal government, to compensate local jurisdictions for tax revenues foregone as a result of ownership by a tax-exempt owner. Any FLP tract acquired in fee and held by the FS is eligible for PILT payments (entitlement land as defined at 31 U.S.C. 6901). Federal funds for PILT are not authorized for any land or interests in land held by nonfederal entities, or for conservation easements held by the United States.

Program Funds are FLP funds that are appropriated by Congress and allocated by the FS to three categories: Project funds, Administrations funds, and AON Preparation funds. Project Evaluation Criteria are developed by the States, in consultation with the State Forest Stewardship Coordinating Committees (SFSCC), to evaluate the eligible tracts submitted by interested landowners for inclusion in the FLP. Relocation refers to the provision in the Uniform Relocation Assistance and Real Estate Property Acquisition Policies Act of 1970 (PL 91-646 or 42 U.S.C. 4601) which requires Federal agencies and programs to pay for the relocation of a person displaced by a federally funded real estate transaction.

Reserved Areas are designated areas where nonforest uses (e.g. house, barn, remote recreation camps, etc.) are or will be allowed, but are inseparable from the land holding and do not have a detrimental effect on the conservation easement values. These areas shall be defined and described in the conservation easement and may be restricted in terms of their use, or provisions made through cost and time to cure and treatment. To the extent possible these areas of non-compliance should be excluded from the FLP project. Reserved Interest Deed is where the grantee (government) acquires all rights, titles, and interests in a property, except those rights, titles, and interests that may run with the land that are expressly reserved by a grantor (landowner).

Secretary is the U.S. Secretary of Agriculture.

State refers to any of the 50 States, Puerto Rico, Guam, the United States Virgin Islands, the Commonwealth of the Mariana Islands, and American Samoa participating in the FLP. State Forest Stewardship Coordinating Committees (SFSCC) are defined, and their duties are described, in Section 19(b) of the CFAA (16 U.S.C. 2113). They are chaired and administered by the State Foresters, or equivalent State officials, with membership composed of representatives from the following agencies, organizations, or individuals: Forest Service; Natural Resources Conservation Service; Farm Services Agency; Cooperative State Research, Education, and Extension Service; local government; consulting foresters; environmental organizations; forest products industry; forest land owners; land trusts; conservation organizations; the State fish and wildlife agency; and others determined appropriate by the Secretary. The SFSCC makes recommendations to the State lead agency regarding the AON, AON amendments, and the determination of project priorities.

State Lead Agency is that unit of State government responsible for coordinating the establishment and implementation of the FLP in the State, as designated by the Governor or pursuant to State law. The State lead agency is usually a forestry agency, but may be another natural resource agency. Tribal Assessments of Need-An AON is developed by a federally recognized Indian Tribe in cooperation with the State and the SFSCC. Only nontrust allotment lands are eligible for FLP. Lands or interests in lands purchased under a Tribal FLP can be through a grant to a cooperating State or through the Federal acquisition option. Working Lands Conservation Committee is a committee of the NASF having coordination and consultation responsibilities within that organization regarding the FLP.

UN's ENVIRONMENTAL CONSCIENCE

UNEP may be regarded as having locus classicus as regards institutionalised international environmental cooperation. It is the product of the United Nations Conference on Human Environment (Stockholm, 1972), which was attended by 113 nations. Many of the States had doubts about that first global effort to address international environmental problems. As such, UNEP became the first major UN environmental institution, in spite of proposals to 'bypass the United Nations in favour of an environmental agency run by the developed countries of the world'. The birth of UNEP at the Stockholm Conference, as a part of the UN system, followed the realisation that 'there are problems in using the UN for environmental action; the problems in bypassing it in favour of other institutions would be more serious still'.

2

ENVIRONMENTAL PROTECTION POLICY

INTRODUCTION

An Environmental Impact Assessment (EIA) is an assessment of the likely positive and/or negative influence a project may have on the environment. "Environmental Impact Assessment can be defined as: The process of identifying, predicting, evaluating and mitigating the biophysical, social, and other relevant effects of development proposals prior to major decisions being taken and commitments made." The purpose of the assessment is to ensure that decision-makers consider environmental impacts before deciding whether to proceed with new projects.

The US Environmental Protection Agency pioneered the use of pathway analysis to determine the likely human health impact of environmental factors. The technology for performing such analysis is properly labelled environmental science.

The principal phenomena or pathways of impact are: soil contamination impacts, air pollution impacts, noise health effects, ecology impacts including endangered species assessment, geological hazards assessment and water pollution impacts. Pathway analysis and The Natural Step definitions subsequently became the basis of the global ISO 14000 series of environmental management standards and the more recent ISO 19011 auditing standard; however, these ISO standards are not in common use in the U.S. and most other countries.

After an EIA analysis, the Precautionary Principle and Polluter Pays may be applied to prevent, limit, or require strict liability or insurance coverage to a project, based on its likely harms.

Environmental impact analysis is sometimes controversial and contested. Related analysis of social impacts is achieved by Social impact assessment. Analysis of business impacts is achieved by context analysis. Design impacts are assessed in relation to context theory. EIA predicts what a specific action can do to the environment.

EIA AROUND THE WORLD

China

The Environmental Impact Assessment Law (EIA Law) requires an environmental impact assessment to be completed prior to project construction. However, if a developer completely ignores this requirement and builds a project without submitting an environmental impact statement, the only penalty is that the environmental protection bureau (EPB) may require the developer to do a make-up environmental assessment. If the developer does not complete this make-up assessment within the designated time, only then is the EPB authorized to fine the developer. Even so, the possible fine is capped at a maximum of about US$25,000, a fraction of the overall cost of most major projects. The lack of more stringent enforcement mechanisms has resulted in a significant percentage of projects not completing legally required environmental impact assessments prior to construction.

China's State Environmental Protection Administration (SEPA) used the legislation to halt 30 projects in 2004, including three hydro-power plants under the Three Gorges Project Company. Although one month later (Note as a point of reference, that the typical EIA for a major project in the USA takes one to two years.), most of the 30 halted projects resumed their construction, reportedly having passed the environmental assessment, the fact that these key projects' construction was ever suspended was notable.

A joint investigation by SEPA and the Ministry of Land and Resources in 2004 showed that 30 to 40 per cent of the mining construction projects went through the procedure of environment impact assessment as required, while in some areas only 6 to 7 per cent did so. This partly explains why China has witnessed so many mining accidents in recent years. SEPA alone cannot guarantee the full enforcement of environmental laws and regulations, observed Professor Wang Canfa, director of the centre to help environmental victims at China University of Political Science and Law. In fact, according to Wang, the rate of China's environmental laws and regulations that are actually enforced is estimated to be barely 10 per cent.

EU

The EIA Directive on Environmental Impact Assessment of the effects of projects on the environment was first introduced in 1985 and was amended in 1997. The directive was amended again in 2003 following the 1998 signature by the EU of the Aarhus Convention on public participation in environmental matters. The issue was enlarged to the assessment of plans and programmes by the so called SEA-Directive in 2001 which is now in force and establishes a mix of mandatory and discretionary procedures for assessing environmental impacts. Under the EU directive, an EIA must provide certain information to comply. There are seven key areas that are required:

Description of the Project

- Description of actual project and site description
- Break the project down into its key components, ie construction, operations, decommissioning
- For each component list all of the sources of environmental disturbance
- For each component all the inputs and outputs must be listed, eg, air pollution, noise, hydrology

Alternatives that have been Considered

- Examine alternatives that have been considered

- Example: in a biomass power station, will the fuel be sourced locally or nationally?

Description of the Environment

- List of all aspects of the environment that may be effected by the development
- Example: populations, fauna, flora, air, soil, water, humans, landscape, cultural heritage
- This section is best carried out with the help of local experts, eg the RSPB in the UK

Description of the Significant Effects on the Environment

- The word significant is crucial here as the definition can vary
- 'Significant' needs to be defined
- The most frequent method used here is use of the Leopold matrix
- The matrix is a tool used in the systematic examination of potential interactions
- Example: in a windfarm development a significant impact may be collisions with birds

Mitigation

- This is where EIA is most useful
- Once section 4 has been completed it will be obvious where the impacts will be greatest
- Using this information ways to avoid negative impacts should be developed
- Best working with the developer with this section as they know the project best
- Using the windfarm example again construction could be out of bird nesting seasons

Non-technical Summary

- The EIA will be in the public domain and be used in the decision making process

- It is important that the information is available to the public
- This section is a summary that does not include jargon or complicated diagrams
- It should be understood by the informed lay-person

Lack of Know-how/technical Difficulties

- This section is to advise any areas of weakness in knowledge
- It can be used to focus areas of future research
- Some developers see the EIA as a starting block for good environmental management

New Zealand

In New Zealand, EIA is usually referred to as Assessment of Environmental Effects (AEE). The first use of EIA's dates back to a Cabinet minute passed in 1974 called Environmental Protection and Enhancement Procedures. This had no legal force and only related to the activities of government departments. When the Resource Management Act was passed in 1991, an EIA was required as part of a resource consent application. Section 88 of the Act spells this out.

United States

Under United States environmental law an EIA is referred to as the Environmental Impact Statement (EIS), and originated in the National Environmental Policy Act (NEPA), enacted in the United States in 1969. Certain actions of federal agencies must be preceded by an EIS. Contrary to a widespread misconception, NEPA does not prohibit the federal government or its licensees/permittees from harming the environment, nor does it specify any penalty if the EIS turns out to be inaccurate, intentionally or otherwise.

NEPA requires that plausible statements as to the prospective impacts be disclosed in advance. The purpose of NEPA process is to ensure that the decision maker is fully informed of the environmental aspects and consequences prior to making the final decision.

Usually, an agency will release a Draft Environmental Impact Statement (DEIS) for comment. Interested parties and the general public have the opportunity to comment on the draft, after which the agency will approve the Final Environmental Impact Statement (FEIS). Occasionally, the agency will later release a Supplemental Environmental Impact Statement (SEIS), especially if environmental conditions suddenly change after the issuance of the FEIS.

The adequacy of an EIS can be challenged in court. Major proposed projects have been blocked because of an agency's failure to prepare an acceptable EIS. One prominent example was the Westway landfill and highway development in and along the Hudson River in New York City.

Another prominent case involved the Sierra Club suing the Nevada Department of Transportation over its denial of Sierra Club's request to issue a supplemental EIS addressing air emissions of particulate matter and hazardous air pollutants in the case of widening US Highway 95 through Las Vegas. The case reached the 9th Circuit Court of the United States, which led to construction on the highway being halted until the court's final decision. The case was settled prior to the court's final decision.

Several US state governments that have adopted "little NEPA's," i.e., state laws imposing EIS requirements for particular state actions and some of those state laws refer to the required environmental impact studies as Environmental Impact Reports or Environmental Impact Assessments. For example, the California Environmental Quality Act (CEQA) requires an Environmental Impact Report (EIR).

These various state requirements are yielding voluminous data not just upon impacts of individual projects, but also to elucidate scientific areas that had not been sufficiently researched. For example, in a seemingly routine Environmental Impact Report for the city of Monterey, California, information came to light that led to the official federal endangered species listing of Hickman's potentilla, a rare coastal wildflower.

ENVIRONMENTAL IMPACT ASSESSMENT OF FORESTRY PROJECTS

The current Regulations are The Environmental Impact Assessment (Forestry) (England and Wales) Regulations 1999 [SI 1999/2228] and the Environmental Impact Assessment (Forestry) (Scotland) Regulations 1999 [SI 1999/43]. These Regulations have been further amended by

The Environmental Impact Assessment (Forestry) (England and Wales) (Amendment) Regulations 2006 and The Environmental Impact Assessment (Scotland) Amendment Regulations 2006. These Regulations came into force on 6th September 1999 and require anyone who wishes to carry out a relevant project (i.e. deforestation, afforestation, forestry roads or quarries that might have a significant effect on the environment) to obtain consent for the work from the Forestry Commission. The applicant or proposer must submit an Environmental Statement in support of the proposals to apply for consent.

You will find below some of the terms used in the EIA Regulations:

a) Appropriate Authority-The Secretary of State for Environment, Food and Rural Affairs in England, the National Assembly for Wales, and the Scottish Executive in Scotland.

b) Countryside bodies

   England-Natural England, Environment Agency. Wales-Countryside Council for Wales, Environment Agency. Scotland-Scottish Natural Heritage, Scottish Environmental Protection Agency. And any other body designated by statutory provision as having specific environmental responsibilities.

c) Determination-Taken from Regulation 15 "Determination of applications" and is the process by which we make our decision about the application for consent.

d) Forestry projects-the project work that the Forestry Commission must assess under these Regulations are deforestation (conversion to another land use), afforestation, forestry roads and forestry quarries.

e) Opinion-Our consideration of the proposals from which we will decide whether or not the project is a relevant one. If it is, the applicant must apply for consent and provide an ES.

f) Relevant project-A forestry project (afforestation, deforestation, forest roads works and forest quarry works) that is likely, by virtue of factors such as its nature, size and location, to have a significant effect on the environment and as such requires the FC's consent.

g) Scoping-A gathering of all consultees and other interested parties to discuss and agree the significant issues of concern that require to be addressed by an applicant when preparing an Environmental Statement.

h) Screening-the process by which the Forestry Commission decides whether a project "is likely to have significant effects on the environment by virtue, inter alia, of its nature, size or location". This process is a distinct one from scoping.

i) Publicity-The Forestry Commission maintains a web-based EIA Register that gives details of all the decisions we make under these Regulations. Applicants are also required to advertise details of any application for consent.

j) Thresholds-Area limits set by the Regulations below which it is not expected that the project will have a significant effect on the environment.

The Forestry Commission will assess whether:

- o the proposed project is one of the following categories-afforestation, deforestation, forest roads or quarries;
- o the area is above the relevant threshold (includes extensions to similar areas of work);

o the project is likely to have a significant effect on the environment.

Our consent will be required to proceed with the work if your proposals meet each of these requirements. You will be asked to provide an Environmental Statement to allow us to decide whether to give consent to the project. Under the 1999 Regulations, proposals are considered to be relevant projects (i.e. to require an EIA) if they fall within the categories listed below and the work proposed is likely to have a significant effect on the environment.

a) Initial afforestation: creating new woods and forests by planting trees (on an area that has not had trees for many years). This category includes using direct seeding or natural regeneration, planting Christmas trees and short rotation coppice;

b) Deforestation: conversion of woodland to another type of land use (e.g.heathland);

c) Forest roads: constructing forestry roads, including those within a forest and those leading to a forest;

d) Forest quarries: quarrying to obtain material (rock, sand and gravel) for the formation, alteration or maintenance of forest roads.

We may serve you with an Enforcement Notice if you carry out work on a project that would have required our consent. This notice will require you to comply with the Regulations. Situations may arise where one of the above forestry projects forms part of a wider development that requires Planning Permission.

In these circumstances, any necessary EIA will usually not be dealt with under the Forestry EIA Regulations but under the parallel Town and Country Planning EIA Regulations.

Where a project has been completed since 6 September 1999, work proposals of the same type (including that on land in different ownerships) that extend it beyond the thresholds may need our consent. This also applies where the accumulated area of all adjacent existing projects plus the new proposals, exceeds the thresholds.

Making the decision Definition of relevant projects Proposals that extend a project area previously below the area thresholds

If the area of the work you are proposing falls below these thresholds then it is not likely to be a relevant project and our consent under the EIA Regulations will not be required. You may still of course require a felling licence or to get approval under the Woodland Grant Scheme for the work.

If your project is above these area thresholds, our consent for carrying out the work may be required.

You must provide the following so that we can give our opinion on whether the work you are proposing is a relevant project and requires our consent:

a) a map identifying the area and showing the extent of the project. (This should be a clear Ordnance Survey map at a scale of 1:10 000 or 1:2 500);

b) a brief description of the proposed work and its possible significant effects on the environment;

c) any other information that might be relevant, e.g. species maps, plans, photographs etc.

In the majority of instances this information will be provided as part of a Woodland Grant Scheme application (initial afforestation) or an application for a felling licence (deforestation) or Forest Plan.

If, at any time, we become aware of proposals that require our consent, we will give our opinion to the person whom we believe should have asked us. We will give our opinion, in writing, within 28 days of receiving the necessary information. However, if we need more information from you to help us decide, then we will let you have a decision within 28 days of receiving this information. If our decision will take longer we will seek to agree an appropriate period with you. If we tell you that your proposal is a relevant project, or we do not give you our opinion within the agreed timescale, you can appeal in writing to: The Scottish Ministers (in relation to Scotland); The Secretary of State for Environment, Food and Rural

Affairs; The National Assembly for Wales in Wales for a "direction" as to whether the work is a relevant project. Once we have told you our decision we will publish this on an "EIA Public Register" held at Forestry Commission offices. This will show information about our opinion whether a project is a relevant project, our decision about the granting of consent and any directions that may have been made by the appropriate authority or the Scottish Ministers.

Getting our Consent

We must have the following information so that we can make a decision:-

a) a map showing the area where the project is proposed. (This should be a clear Ordnance Survey map at a scale of 1:10 000 or 1:2 500);

b) a brief description of the proposed work and the possible significant effects on the environment;

c) any other information that might be relevant e.g. species maps, plans, photographs etc.

(You may already have supplied sufficient information for the above if you have asked for our Opinion under the EIA regulations or submitted a Woodland Grant Scheme or Forest Plan application)

d) an Environmental Statement for the work;

e) a copy of the publicity notice that you must place in newspapers.

Before you start to prepare your ES, we would encourage you to meet us and discuss the significant environmental issues that will need to be covered, or ask for information to help you. At an early stage, it is strongly recommended that you contact the appropriate countryside bodies* and request any information they may have that is relevant to the project. [Please note that there may be a small charge for providing this information].

* In England-Natural England, English Heritage, Environment Agency; In Wales-Countryside Council for Wales, Environment Agency;

In Scotland-Scottish Natural Heritage, Scottish Environmental Protection Agency

In addition we would advise that you hold a scoping meeting with countryside bodies and other relevant organisations to help determine the issues that must be covered in the Environmental Statement. There is no set format for an ES but it must contain the standard information contained in Appendix 2. Your Statement must include everything from the list that relates to your application, placing the emphasis on the significant environmental effects. You can also get more advice about preparing an Environmental Statement in our booklet-"Preparing an Environmental Statement for a Forestry Project", only available as a PDF download from www.forestry.gov.uk.

It may help to facilitate your application if you let us have a draft copy of your ES before preparing the final copy. This will help us to decide if you have covered and addressed all the relevant issues. If we need more information to help us decide about your application, you must supply this. You can get a copy of this booklet on the Forestry Commission Grants and Licences section of the website. If you have applied for our consent and prepared an ES, you must publish a notice giving the details outlined below. We will agree at least two newspapers in which the notice must appear.

The notice must:

a) state that an application for our consent under the Regulations has been made;

b) indicate where (and at what times) copies of the application for consent can be inspected (the local Forestry Commission office or other location(s) we agree);

c) state the address where people can get copies of the application;

d) give details of any charge that may be made for copies-to cover printing, copying and distribution costs only;

e) state that comments about the application must be made to the relevant Forestry Commission Conservator within 28 days of the date of the notice.

f) You must also include the following statement: "The Forestry Commissioners may decide either to grant consent subject to the standard conditions required by Regulation 18 of the EIA Regulations, or subject to such further conditions as they see fit or refuse consent."

We will also give details of your application to the appropriate countryside body as well as the local authority with an interest in the application. They are required to give us their comments within 28 days.

We will give you our decision (and the reasons for it) in writing, once the 28-day period for comments has expired and we have considered the environmental information, any representations and any other material consideration. A notice giving details of our decision will be published in the same newspapers in which the initial notice about the application appeared. We will also inform anyone from whom we have received representations in relation to the application.

When granting consent, we will specify the date by which you must start and finish the work. The start date will be no later than five years from the date of the consent and the finish date will be no later than ten years from the date of the consent. We may also apply other conditions to the consent that we deem necessary to ensure that the environment is protected from the impact of the project. These may be in addition to the normal Grant Scheme or Felling Licence conditions. Publicity about the application Consultation about the application Giving our decision about your application Standard consent conditions.

If you have applied for consent, you can appeal about the decision we make to:

*The Scottish Ministers (in relation to Scotland);

*The Secretary of State for Environment, Food and Rural Affairs in England;

*The National Assembly for Wales in Wales;

You must do this within 28 calendar days of receiving notice of our decision or within any other period that the Government authority* may agree.

You may only appeal if the Forestry Commissioners have:

a) refused the application;

or

b) have granted consent with conditions (additional to the standard conditions); or

c) we have asked you to start and finish the work within a period that is less than the maximum period permitted by the Regulations. In addition, any person who feels aggrieved by the granting of consent can make an application to the Court if they feel that their interests are prejudiced. They must do this within 6 weeks of consent being granted.

If we discover that you have carried out work that would have required our consent, or have breached the conditions of consent, we may serve you with an Enforcement Notice. This will require you to do one or more of the following:

a) stop the work;

b) apply for consent;

c) restore the land to its condition before the work was started;

d) carry out work to secure compliance with the conditions of the consent;

e) remove or alleviate any damage to the environment that has been caused by the work.

You can appeal against the serving of the Enforcement Notice to:

*The Scottish Ministers (in relation to Scotland);

*The Secretary of State for Environment, Food and Rural Affairs in England;

*The National Assembly for Wales in Wales;

You must do this within 28 calendar days of receiving notice of our decision or within any other period that the Government authority* may agree.

If you require more information or help, including how to make appeals, contact your local Forestry Commission office or: Grants and Licences, Forestry Commission Headquarters, 231 Corstorphine Road, Edinburgh EH12 7AT Appealing against our decision Carrying out work without our consent or breaching the conditions of consent Appealing against the enforcement notice Getting more information.

Definition of "sensitive areas":

For the purposes of the the definitions of "sensitive areas" are as follows:

a) National Nature Reserve or Site of Special Scientific Interest;

b) National Parks;

c) The Broads;

d) A World Heritage Site;

e) Scheduled Ancient Monuments;

f) Area of Outstanding Natural Beauty;

g) Natural Heritage or National Scenic Areas;

h) A site designated or identified as a candidate Special Area of Conservation;

i) The New Forest Heritage Area;

j) A site classified or proposed as a Special Protection Area.

Information for inclusion in environmental statements Part 1.

1) Description of the project, including in particular:

   a) a description of the physical characteristics of the whole project and the land-use requirements during the construction and operational phases;

   b) a description of the main characteristics of the production processes, for instance, nature and quanuty of the materials used;

c) an estimate, by type and quantity, of expected residues and emissions (water, air and soil pollution, noise, vibration, light heat, radiation, etc) resulting from the operation of the proposed project.

2) An outline of the main alternatives studied by the developer and an indication of the main reasons for the developer's choice, taking into account environmental effects.

3) A description of the aspects of the environment likely to be significantly affected by the proposed project, including, in particular, human beings, fauna, flora, soil, water, air, climatic factors, material assets including the architectural and archaeological heritage, landscape and the inter-relationship between the above factors.

4) A description of the likely significant effects of the proposed project on the environment, which should cover the direct effects and any indirect, secondary, cumulative, short, medium and long-term, permanent and temporary, positive and negative effects of the project, resulting from:

a) the existence of the project;

b) the use of natural resources;

c) the emission of pollutants, the creation of nuisances, and the elimination of waste, and the description by the developer of the forecasting methods used to assess the effects on the environment.

5) A description of the measures proposed to prevent, reduce and, where possible offset, any significant adverse effects on the environment.

6) A non-technical summary of the information provided under paragraphs 1 to 5 above.

7) An indication of any difficulties (technical deficiencies or lack of knowledge) encountered by the developer in compiling the required information.

GUIDE TO PROCEDURES

Environmental impact assessment (EIA) is an important procedure for ensuring that the likely effects of new development on the environment are fully understood and taken into account before the development is allowed to go ahead. This booklet, which is intended primarily for developers and their advisers, explains how European Community (EC) requirements for the environmental impact assessment of major projects have been incorporated into consent procedures in the UK. It revises the booklet 'Environmental Assessment: A Guide to the Procedures', first published in 1989, to take account of the requirements of Directive 97/11/EC, which was adopted on 3 March 1997 and came into effect on 14 March 1999. Directive 97/11/EC amends the original Directive 85/337/EEC on 'The assessment of the effects of certain public and private projects on the environment', which came into effect in July 1988. The text of the Directive as amended (with amendments shown in italics) is reproduced in Appendix 1 to this booklet. Throughout this booklet, references to `the Directive' mean Directive 85/337/EEC as amended by Directive 97/11/EC.

The effect of the Directive is to require environmental impact assessment to be carried out, before development consent is granted, for certain types of major project which are judged likely to have significant environmental effects. Parts 1 and 2 of this booklet explain the procedures which apply to projects falling within the scope of the Directive and requiring planning permission in England and Wales.

For such projects the Directive was given legal effect through the Town and Country Planning which came into force on 14 March 1999 and apply to relevant planning applications lodged on or after that date. The full text of the Regulations is published separately by The Stationery Office. For ease of reference, those parts of the Regulations which list the types of project to which they apply, and specify what information an environmental statement must contain, are reproduced in Appendices 2 to 4 to this booklet. The booklet is not intended to be an authoritative interpretation of the law and does not remove the need to refer to the Regulations.

Formal guidance on procedures under the Regulations, directed principally at local planning authorities, was issued in DETR Circular 2/99 (Welsh Office Circular 11/99). Although the present booklet, like its predecessor, is meant to be reasonably self-contained, developers may need to refer to that Circular, particularly for fuller information on how planning authorities are expected to judge the significance of a project's likely effects for the purpose of deciding whether environmental impact assessment is required. Parts 1 and 2 also give some general guidance, applicable to all types of project, on the nature of environmental impact assessment and on the practical aspects of preparing an environmental statement. Part 3 gives a brief account of the procedures which apply to other projects within the scope of the Directive but which are not approved under planning procedures, for example, motorways, harbour works and long distance pipe-lines. It also deals very briefly with environmental impact assessment procedures in Scotland and Northern Ireland. For the detailed requirements, reference will need to be made to the relevant statutory instruments and associated guidance.

Throughout this booklet, the term 'environmental impact assessment' (EIA) is used to describe the whole process whereby information about the environmental effects of a project is collected, assessed and taken into account in reaching a decision on whether the project should go ahead or not. The process was formerly referred to in the UK as 'environmental assessment' (EA). An 'environmental statement' is a document setting out the developer's own assessment of a project's likely environmental effects, which is prepared and submitted by the developer in conjunction with the application for consent. Go to table of contents

ENVIRONMENTAL IMPACT ASSESSMENT AND PROJECTS WHICH REQUIRE PLANNING PERMISSION

The term 'environmental impact assessment' (EIA) describes a procedure that must be followed for certain types of project before they can be given 'development consent'. The procedure is a means of drawing together, in a systematic way, an assessment of a project's likely significant environmental effects. This helps to ensure that the importance of the predicted effects, and the scope for reducing them, are properly understood by the public and the relevant competent authority before it makes its decision.

Environmental impact assessment enables environmental factors to be given due weight, along with economic or social factors, when planning applications are being considered. It helps to promote a sustainable pattern of physical development and land and property use in cities, towns and the countryside. If properly carried out, it benefits all those involved in the planning process.

From the developer's point of view, the preparation of an environmental statement in parallel with project design provides a useful framework within which environmental considerations and design development can interact. Environmental analysis may indicate ways in which the project can be modified to avoid possible adverse effects, for example through considering more environmentally friendly alternatives. Taking these steps is likely to make the formal planning approval stages run more smoothly.

For the planning authority and other public bodies with environmental responsibilities, environmental impact assessment provides a basis for better decision making. More thorough analysis of the implications of a new project before a planning application is made, and the provision of more comprehensive information with the application, should enable authorities to make swifter decisions.

While the responsibility for compiling the environmental statement rests with the developer, it is expected that the developer will consult those with relevant information, and the Regulations specifically require that public authorities which have information in their possession which is relevant to the preparation of the environmental statement should make it available to the developer.

The general public's interest in a major project is often expressed as concern about the possibility of unknown or unforeseen effects. By providing a full analysis of a project's effects, an environmental statement can help to allay fears created by lack of information. At the same time, early engagement with the public when plans are still fluid can enable developers to make adjustments which will help to secure a smoother passage for the proposed development and result in a better environmental outcome. The environmental statement can also help to inform the public on the substantive issues which the local planning authority will have to consider in reaching a decision.

It is a requirement of the Regulations that the environmental statement must include a description of the project and its likely effects together with a summary in non-technical language. One of the aims of a good environmental statement should be to enable readers to understand for themselves how its conclusions have been reached, and to form their own judgements on the significance of the environmental issues raised by the project.

Environmental impact assessment can therefore be helpful to all those concerned with major projects. The following paragraphs describe the procedures for deciding whether EIA is necessary in a particular case and, where it is, for carrying out the assessment. The procedure is intended to make the most of the potential benefits of EIA, while keeping the process as simple and flexible as possible, and avoiding any duplication of existing planning procedures.

NEED OF ENVIRONMENTAL IMPACT ASSESSMENT

The Regulations apply to two separate lists of projects:

i. 'Schedule 1 projects', for which EIA is required in every case;
ii. 'Schedule 2 projects', for which EIA is required only if the particular project in question is judged likely to give rise to significant environmental effects. Lists of Schedule 1 and Schedule 2 projects are given in, respectively, Appendices 2 and 3 to this booklet.

For Schedule 1 projects, whether or not a particular project falls within the scope of the Regulations will normally be clear: several of the definitions of Schedule 1 projects incorporate an indication of scale, in the form of a quantified threshold, which clearly identifies the projects requiring EIA. Where there is any doubt about a project's inclusion in Schedule 1, the procedures described in below can be used to obtain an opinion from the planning authority or a direction from the Secretary of State (or, in Wales, the National Assembly for Wales).

For the much longer list of Schedule 2 projects, the issue turns on the likelihood of 'significant environmental effects'. For the different types of project described in column 1 of Schedule 2, the 1999Regulations introduced a system of thresholds and criteria, shown in column 2,as a method of discounting development which is not likely to have significant effects on the environment. For development where the applicable threshold or criterion is not exceeded or met, EIA is not normally required.

However, even where the threshold or criterion is not met or exceeded, EIA may be required if the proposed development is in, or partly in, a 'sensitive area.' In exceptional circumstances the Secretary of State (or, in Wales, the National Assembly for Wales) may exercise his power under the Regulations to direct that a particular type of Schedule 2 development requires EIA even if it is not to be located in a sensitive area and does not exceed or meet the applicable threshold or criterion.

The more environmentally sensitive the location, the more likely it is that the effects of development will be significant and that EIA will be required. That is why the thresholds and criteria do not apply where development is proposed in, or partly in, a `sensitive area' as defined in the Regulations. Such areas include Sites of Special Scientific Interest (SSSIs), National Parks, Areas of Outstanding Natural Beauty, the Broads, World Heritage Sites and scheduled monuments. There is no general presumption that every Schedule 2 development in a sensitive area will require EIA. Nevertheless, in the case of development to be located in or close to SSSIs, especially those which are also international conservation sites such as Ramsar sites or Special Protection Areas for birds, the likely environmental effects will often be such as to require EIA.

HOW 'SIGNIFICANCE' WILL BE ASSESSED

Developments which meet or exceed the applicable threshold are considered on a caseby-case basis. For the purpose of determining whether EIA is necessary, those of the selection criteria set out in Schedule 3 to the Regulations which are relevant to the proposed development, must be taken into account. The selection criteria fall into the three broad headings: characteristics of the development, location of the development, and characteristics of the potential impact.

For obvious reasons there can be no general definition of what constitutes significance.

General guidance on how to assess `significance' is contained in DETR Circular 2/99 (Welsh Office Circular 11/99);and rulings may be obtained from the local planning authority or the Secretary of State (or, in Wales, the National Assembly for Wales) on whether EIA is required in particular cases. Essentially the Circular suggests that there are three main criteria of significance:

i. major developments which are of more than local importance;
ii. developments which are proposed for particularly environmentally sensitive or vulnerable locations;
iii. developments with unusually complex and potentially hazardous environmental effects.

These are very general guidelines and, to assist in their application to particular cases, the Circular also sets out indicative thresholds and criteria by reference to particular categories of development listed in Schedule 2 to the Regulations. These are reproduced in the last column of the Table in Appendix 3 to this booklet. It will be obvious that none of these guidelines can be applied as hard and fast rules; circumstances are bound to vary greatly from case to case.

Some large-scale projects which exceed the indicative thresholds may not be significant enough to require EIA; some smaller projects, particularly in sensitive locations, may be candidates for EIA. Nevertheless, the guidance in the Circular should provide a starting point for consideration by the developer and the planning authority of the need for EIA. If the matter is referred to the Secretary of State (or, in Wales, the National Assembly for Wales), he will have regard to the published criteria.

OBTAINING A RULING ON THE NEED FOR EIA

Where there is a possibility that a proposed development will require environmental impact assessment, developers are advised to consult the relevant planning authority well in advance of a planning application. Developers can decide for themselves that a given project falls within the scope of the Regulations so that an environmental statement will be needed.

But the Regulations also provide a procedure which enables developers to apply to the planning authority for an opinion ('screening opinion') on whether EIA is needed in a particular case, as soon as a basic minimum of information can be provided about the proposal. This must include a plan on which the site of the proposed development is identified, and a brief description of its nature and purpose and of its possible effects on the environment. This can, of course, be supplemented with other information if the developer wishes.

Where such information can be provided, the developer may approach the planning authority at any time for an opinion on the need for EIA. This can be done well in advance of any formal planning application, although any approach to the planning authority before the planning application stage is entirely voluntary. Where such an approach is made, the planning authority must give its opinion within three weeks, unless the developer agrees to a longer period. The planning authority may request further information from the developer, but this in itself does not extend the three-week time limit, unless the developer agrees. The planning authority must make its determination available for public inspection at the place where the appropriate register (or relevant section of that register) is kept.

Where the planning authority expresses the opinion that a particular proposal requires EIA-whether in response to a request from a developer prior to a planning application, or following a planning application-it must provide a written statement giving clear and precise reasons for its opinion. Both that statement and the developer's application for an opinion are then made available for public inspection at the same place as the register.

A developer who is dissatisfied with the planning authority's opinion that EIA is required may refer the matter to the Secretary of State (or, in Wales, the National Assembly for Wales). The developer is simply required to copy the relevant papers to the Government Office in the region concerned (or the Assembly) and add any representations which are considered to be appropriate in the light of the planning authority's statement.

The Secretary of State (or the Assembly) will then normally give a direction within three weeks of the developer's application; and, if the direction is to the effect that EIA is required, it will be accompanied by a statement of reasons. The developer may, when requesting a screening opinion from the planning authority, simultaneously request an opinion on what should be included in any environmental statement.

The broad intention of this procedure is to ensure that developers can obtain a clear ruling on the need for EIA well before they reach the stage of lodging a formal planning application. This should minimise the possibility of delay or uncertainty at that stage. Where the matter is not raised until a formal planning application is lodged, the developer risks serious delay if either the planning authority or the Secretary of State (or, in Wales, the National Assembly for Wales) rules that an environmental statement must be prepared. No action will be taken on the planning application until the developer has prepared an environmental statement and submitted it to the planning authority.

In most cases this procedure will give developers a firm decision on the need for EIA as soon as they can provide basic information about their project. There may occasionally be cases where the receipt of information directly leads the Secretary of State (or, in Wales, the National Assembly for Wales) to issue a direction when there has been no request for one from the developer; or to overrule a planning authority's opinion that EIA is not required; or even, very exceptionally, to reverse an earlier direction. This could, for instance, happen where a development proposal came to the Secretary of State's (or the Assembly's) notice on 'call-in'; or where third party representations drew attention to aspects of a proposed development which were not known to the Secretary of State (or the Assembly) when the initial direction was given. Exceptionally, also, a planning authority which has given a pre-application opinion that EIA is not required might consider it necessary to reverse that decision when the planning application is formally submitted. In that case the developer could, of course, apply to the Secretary of State (or the Assembly) for a direction.

'PERMITTED DEVELOPMENT RIGHTS' (PDRS)

Developments which do not require planning permission because of the provisions of the Town and Country Planning (General Permitted Development) PDRs provided that they do not fall into Schedule 1 or 2 of the EIA Regulations. For developments that do fall within Schedule 1 or 2, the general position is as follows. Schedule 1projects are not permitted development, and always require the submission of a planning application and an environmental statement. PDRs for Schedule 2 projects which either exceed or meet the applicable threshold or criterion, or are wholly or partly in a sensitive area, are also withdrawn, unless the local planning authority has adopted a screening opinion (or the Secretary of State (or, in Wales, the National Assembly for Wales) has directed) to the effect that EIA is not required. There are exceptions to these provisions in the case of the following classes in Schedule 2 to the 1995 Order: Part 7, Class D of Part 8, Part 11, Class B of Part 12, Class F (a) of Part 17, Class A of Part 20, Class B of Part 20, and Class B of Part

These exceptions exist for a variety of reasons. For example, some relate to projects subject to alternative consent procedures, and others to projects begun before Directive 85/337/EEC came into operation. Further information about the exceptions, and about permitted development and EIA generally, can be found in DETR Circular 2/99 (Welsh Office Circular 11/99).

SIMPLIFIED PLANNING ZONES AND ENTERPRISE ZONES

Special considerations apply to projects proposed for simplified planning zones (SPZs) and enterprise zones (EZs).

Simplified Planning Zones

All Schedule 1 projects are excluded from the scope of SPZs and therefore require EIA as part of the application for planning permission. Where the terms of SPZ schemes would permit Schedule 2 projects to be undertaken without specific planning permission, developers must notify the planning authority that they intend to undertake such development. This notification will give the planning authority and, where appropriate, the Secretary of State (or, in Wales, the National Assembly for Wales) the opportunity to consider whether the project is likely to give rise to significant environmental effects. If so, planning permission will be required in the normal way and an environmental statement must be prepared. If not, the project will enjoy the benefit of the general permission granted by the SPZ scheme, and no separate application for planning permission will be necessary.

Enterprise Zones

If the planning scheme allows for Schedule 1 projects, EIA must be carried out. As regards Schedule 2 projects, the same procedures apply as those for SPZs described above.

The arrangements in SPZs and EZs are explained in DETR Circular 2/99 (Welsh Office Circular 11/99).

Go to table of contents

Preparing an Environmental Statement: the Planning Procedures

- Preliminary consultations
- Content of the environmental statement
- Statutory and other consultees; the general public
- Techniques of assessment; sources of advice
- Submission of environmental statement with planning application
- Handling by the planning authority
- Requests for further information

- Determination of application
- Appeals and call-ins
- Procedural stages

The developer is responsible for preparing the environmental statement which must be submitted with the planning application. A developer may choose to engage consultants for some or all of the work. The preparation of the statement should be a collaborative exercise involving discussions with the local planning authority, statutory consultees and possibly other bodies as well. There is no prescribed form of statement, provided that the requirements of the Regulations are met. The Regulations enable a developer, before making a planning application, to ask the local planning authority for its formal opinion ('scoping opinion') on the information to be included in an environmental statement. The developer can therefore be clear as to what the authority considers are the aspects of the environment which would be affected. The request may be made at the same time as for the screening opinion, and must be accompanied by the same information.as for the screening opinion. A developer may also wish to submit a draft outline of the environmental statement, indicating what the developer thinks the main issues are, as a focus for the authority's consideration. The authority must consult certain bodies and the developer before adopting a scoping opinion, and must adopt the scoping opinion within five weeks of receiving the request. This period may be extended provided that the developer agrees. The scoping opinion must be kept available for public inspection for two years, with the request (including documents submitted by the developer as part of that request), at the place where the planning register is kept.

There is no provision for an appeal to the Secretary of State (or, in Wales, the National Assembly for Wales) if the developer and local planning authority disagree about the content of an environmental statement. If an authority fails to adopt a scoping opinion within five weeks (or any agreed extension), the developer may apply to the Secretary of State (or the Assembly) for a scoping direction. The application must be accompanied by the previous documents relating to the request for a scoping opinion, plus any additional representations the developer wishes to make. The developer should also send a copy of the request and any representations to the authority. The Secretary of State (or the Assembly) must make a scoping direction within five weeks of receiving the request, or such longer period as he may reasonably require. He must consult the developer and certain bodies beforehand. A copy of the scoping direction will be sent to the developer and the authority.

The aim should be to provide as systematic and objective an account as is possible of the significant environmental effects to which the project is likely to give rise. Where the statement embodies or summarises the conclusions of more detailed work, sufficient information should be provided to enable those who wish to do so to verify the statement's conclusions and to identify the source of the information provided. The environmental statement must contain a non-technical summary which will enable non-experts to understand its findings.

Preliminary Consultations

One of the main emphases of the process of environmental impact assessment is on the need for fuller and earlier consultation by the developer with bodies which have an interest in the likely environmental effects of the development proposal. If important issues are not considered at a very early stage, they may well emerge when a project's design is well advanced, and necessitate rethinking and delay. Ideally, EIA should start at the stage of site selection and (where relevant) process selection, so that the environmental merits of practicable alternatives can be properly considered.

While a developer is under no formal obligation to consult about the proposal before the submission of a formal planning application, there are good practical reasons for doing so. Authorities will often possess useful local and specialised information which is relevant to a project's design, and officers may be able to give preliminary advice about local problems and about those aspects of the proposal that are likely to be of particular concern to the authority.

The timing of such informal consultations is at the developer's discretion; but it will generally be advantageous for them to take place as soon as the developer is in a position to provide sufficient information about the proposal to form a basis for discussion. The developer can ask that any information provided at this preliminary stage should be treated in confidence by the planning authority and any other consultees. If the developer is seeking a formal opinion from the planning authority on the need for environmental impact assessment, the information about the project which accompanies that request will be made public by the authority.

Content of the Environmental Statement

Developers and authorities should discuss the scope of an environmental statement before its preparation is begun. The formal requirements as to the content of environmental statements are set out in Schedule 4 to the Regulations, which is reproduced in Appendix 4 to this booklet. The information is given under separate headings: Part I and Part II. The statement must include at least the information included in Part II, and such of the information in Part I as is reasonably required to assess the environmental effects of the development and which the applicant can reasonably be required to complete. As a practical guide to the range of issues which may need to be considered, developers may find it helpful to use the checklist at Appendix 5 to this booklet as a basis for their discussions with the planning authority. The checklist is not meant to be regarded as a prescribed framework for all environmental statements. Its main purpose is to act as a guide or agenda for the preliminary discussions about the scope of the statement.

The comprehensive nature of the checklist at Appendix 5 should not be taken to imply that all environmental statements should cover every conceivable aspect of a project's potential environmental effects at the same level of detail. They should be tailored to the nature of the project and its likely effects. Whilst every environmental statement should provide a full factual description of the project, the emphasis of Schedule 4 is on the main or significant effects to which a project is likely to give rise.

In some cases, only a few of the aspects set out in the checklist will be significant in this sense and will need to be discussed in the statement in any great depth. Other issues may be of little or no significance for the particular project in question, and will need only very brief treatment, to indicate that their possible relevance has been considered.

It should be noted that developers are now required to include in the environmental statement an outline of the main alternative approaches to the proposed development that they may have considered, and the main reasons for their choice. It is widely regarded as good practice to consider alternatives, as it results in a more robust application for planning permission. Also, the nature of certain developments and their location may make the consideration of alternatives a material consideration. Where alternatives are considered, the main ones must be outlined in the environmental statement.

Even where a local planning authority has adopted a scoping opinion, the developer is responsible for the content of the statement which is finally submitted. Developers should bear in mind that planning authorities have powers to call for additional information when considering environmental statements and planning applications, and that they are likely to use those powers if they consider that aspects of a submitted environmental statement are inadequate. There is no provision for any disagreement between the developer and the planning authority over the content of an environmental statement to be referred to the Secretary of State (or, in Wales, the National Assembly for Wales), except through normal planning appeal procedures.

Developers should consider at an early stage whether an assessment of environmental effects may also be required under another European Community Directive, such as the Habitats Directive (92/43/EEC), the Wild Birds Directive(79/409/EEC), the Integrated Pollution Prevention and Control Directive(96/61/EC) or the Control of Major Accident Hazards Directive (96/82/EC).Although the requirements of these and of the EIA Directive are all independent of each other, there are clearly links between them. Where more than one regime applies, developers could save unnecessary time and effort if they identify and co-ordinate the different assessments required. Advice on the links between the EIA system and the requirements of the Habitats Regulations is offered in PPG 9on Nature Conservation (or, in Wales, Planning Guidance (Wales) Planning Policy First Revision), and on the links between the Town and Country Planning system and the IPPC authorisation system in PPG 23 on Planning and Pollution Control (or, in Wales, Planning Guidance (Wales) Planning Policy First Revision and Planning Guidance (Wales) Technical Advice Note (Wales) 5 'Nature Conservation and Planning').

STATUTORY AND OTHER CONSULTEES; THE GENERAL PUBLIC

The Regulations give a particular role in environmental impact assessment to those public bodies with statutory environmental responsibilities who must be consulted by the planning authority before a Schedule I or a Schedule 2 planning application is determined. A full list of these statutory consultation bodies is given in Appendix 6 to this booklet.

Where the planning authority or the Secretary of State(or, in Wales, the National Assembly for Wales) rules that EIA is required, those bodies which are statutory consultees for the particular project in question will be notified and the developer will be informed accordingly.

The effect of this notification is to put those bodies under an obligation to provide the developer (on request) with any information in their possession which is likely to be relevant to the preparation of the environmental statement. An example might be information held by English Nature (or, in Wales, the Countryside Council for Wales) about the ecology of a particular area, which could be relevant to the assessment of a project's likely effects.

It is up to the developer to approach the statutory consultees and indicate what sort of information would be helpful in preparing the environmental statement. The obligation on statutory consultees relates only to information already in their possession; they are not required to undertake research on behalf of the developer. Nor, at this stage, would consultees be expected to express a view about the merits of the proposal; their views on merits are invited at a later stage. Consultees may make a reasonable charge to cover the cost of making information requested by a developer available.

Developers should also consider whether to consult the general public, and non-statutory bodies concerned with environmental issues, during the preparation of the environmental statement. Bodies of this kind may have particular knowledge and expertise to offer. Some are national organisations, for instance, the Royal Society for the Protection of Birds; inmost areas there are also active local amenity societies and environmental groups.

While developers are under no obligation to publicise their proposals before submitting a planning application, consultation with local amenity groups and with the general public can be useful in identifying key environmental issues, and may put the developer in a better position to modify the project in ways which would mitigate adverse effects and recognise local environmental concerns. It will also give the developer an early indication of the issues which are likely to be important at the formal application stage if, for instance, the proposal goes to public inquiry.

TECHNIQUES OF ASSESSMENT; SOURCES OF ADVICE

Extensive literature is available on how to assess the effects on the environment of particular processes and activities. The assessment techniques used, and the degree of detail in which any particular subject is treated in an environmental statement, will depend on the character of the proposal, the environment which it is likely to affect, and the information available. While a careful study of the proposed location will generally be needed (including environmental survey information), original scientific research will not normally be necessary. The local planning authority and statutory consultees may be able to advise the developer on sources of specialist information, for example, about particular local conditions.

Environmental statements will often need to recognise that there is some uncertainty attached to the prediction of environmental effects. Where there is uncertainty, it needs to be explicitly recognised. Uncertainty is not in itself a reason for discounting the importance of particular potential environmental effects, simply because other effects can be more confidently predicted.

Submission of Environmental Statement with Planning Application

To enable a planning application to be processed as quickly as possible, it is in the developer's interest to submit an environmental statement at the same time as the application is made. It will be for the planning authority to judge how much information is required in the particular case, but the preparation of an environmental statement is bound to require the developer to work out proposals in some detail; otherwise any thorough appraisal of likely effects will be impossible.

Where an application is in outline, the planning authority will still need to have sufficient information on a project's likely effects to enable it to judge whether the development should take place or not. The information given in the environmental statement will have an important bearing on whether matters may be reserved in an outline permission; it will be important to ensure that the development does not take place in a form which would lead to significantly different effects from those considered at the planning application stage.

When the developer submits an environmental statement at the same time as the planning application, three further copies must also be submitted for onward transmission by the planning authority to the Secretary of State (or, in Wales, the National Assembly for Wales). The developer is also required to provide the planning authority with sufficient copies of the environmental statement to enable one to be sent to each of the statutory consultees. Alternatively, the developer may send copies of the statement directly to the consultees. When submitting the application, the developer must inform the planning authority of the name of every body-whether or not it is a statutory consultee-to which a copy of the statement has been sent.

The developer should make a reasonable number of copies of the statement available for members of the public. A reasonable charge reflecting printing and distribution costs may be made.

HANDLING BY THE PLANNING AUTHORITY

Where the environmental statement is submitted with the planning application, the local planning authority will arrange for a notice to be published in a local newspaper and displayed at or near the site of the proposed development. If the environmental statement is submitted after the planning application, responsibility for publicising it falls to the developer.

The planning authority will place the planning application on Part I of the planning register, together with the environmental statement. The authority and the developer may wish to consider the need for further publicity at this stage, for example, publication of further details of the project in a local newspaper, or an exhibition.

The planning authority will also need to notify statutory consultees of the application (unless the developer has already done so) and invite them to comment on the environmental statement. Consultees must be allowed at least 14 days from receipt of the statement in which to comment before a decision is taken. It will often be useful for the planning authority to discuss the project with consultees who have a particular interest in its environmental effects before reaching its conclusions on the planning applications.

The copies of the environmental statement that are forwarded by the planning authority to the Secretary of State (or, in Wales, the National Assembly for Wales), will assist in monitoring. In those exceptional circumstances where a proposed development is likely to have significant effects on the environment in another country, they will also enable certain international obligations for exchange of information with other countries to be met.

Requests for Further Information

Where the planning authority considers that the information provided in the developer's environmental statement, together with that available to the authority from other sources, is insufficient to permit a proper evaluation of the project's likely environmental effects, the authority can require further information, or evidence to verify the information that has already been provided.

The use of these powers should not normally be necessary, especially if the parties have worked together during the preparation of the environmental statement. Nevertheless, further consultation between the planning authority and the developer may be necessary at this stage, in particular to consider comments made by consultees and, possibly, amendments to the proposal to meet objections that have been raised. Where an authority considers that it does not have the necessary expertise to evaluate the information contained in an environmental statement, it may decide to seek advice from consultants or other suitably qualified persons or organisations.

Determination of Application

The planning authority is required to determine a planning application which is the subject of environmental impact assessment within 16 weeks from the date of receipt of the environmental statement, unless the developer agrees to a longer period. In determining the application, the authority is, of course, required to have regard to the environmental statement, as well as to other material considerations. As with any other planning application, the planning authority may refuse permission or grant it with or without conditions.

The planning authority cannot take the view that a planning application is invalid because it considers that an inadequate environmental statement has been submitted or because the developer has failed to provide any further information required under the powers described above. However, if the developer fails to provide enough information to complete the environmental statement, the application can be determined only by refusal.

Appeals and Call-ins

The right of appeal to the Secretary of State (or, in Wales, the National Assembly for Wales) against an adverse decision by a planning authority (or against an authority's failure to determine an application within the 16-week time limit) is the same for planning applications to which the EIA Regulations apply as for other applications. Similarly, the Secretary of State's (or the Assembly's) power to call in a planning application applies in these cases. Where an environmental statement has been prepared to accompany a planning application, the information which it contains will be among the material considerations which an Inspector will take into account in considering an appeal. The Secretary of State (or the Assembly) and Inspectors, like the planning authority, have power to request the developer to provide further information where they consider that the environmental statement is inadequate as it stands. Any additional information provided by the developer in response to such a request will be made available to all parties to an appeal.

Procedural Stages

Appendix 7 to this booklet provides illustrative flowcharts for the five main procedural stages: application to the planning authority for a screening opinion; application to the Secretary of State (or, in Wales, the National Assembly for Wales) for a screening direction; application to the planning authority for a scoping opinion; application to the Secretary of State (or the Assembly) for a scoping direction; and submission of an environmental statement to the planning authority in conjunction with a planning application.

Arrangements for other Projects

- The trunk road network
- Oil and gas pipe-lines
- Offshore oil and gas projects
- Nuclear power stations
- Other power stations and overhead power lines
- Forestry projects

- Land drainage improvements
- Ports and harbours
- Marine fish farming
- Marine dredging for minerals
- Transport and Works Act 1992
- Projects arising in Scotland and Northern Ireland

However, those Parts specifically describe the arrangements for EIA of projects in England and Wales which are approved through the planning system. Part 3 gives brief guidance on the legislative provisions on EIA for projects insofar as they are outside the scope of the planning system.

For all the following projects the Regulations described, which implement the Directive, include provisions for seeking a scoping opinion from the competent authority, for consulting statutory and other bodies, for the submission of an environmental statement in the prescribed form, and for publicising the statement. Where applicable, the Regulations disapply any thresholds for Annex II projects which are at least partly in a 'sensitive area', this term being defined in the same or almost the same way as in the planning EIA Regulations.

PROJECTS NOT SUBJECT TO PLANNING CONTROL

The Trunk Road Network

EIA is mandatory for projects to construct new motorways and certain other roads, including those with four or more lanes, and for certain road improvements.

Roads in England and Wales for which the Secretary of State for Transport or the National Assembly for Wales is the highway authority, are approved under procedures set out in the Highways Act 1980. These roads-the trunk road network-currently include nearly all motorways. The Department concerned will normally consult the public widely about alternative routes before selecting a preferred route for a new road. Once the preferred route has been announced, detailed design work is carried out leading to the publication of statutory orders. In most cases these orders are the subject of a public inquiry held by an independent inspector.

For these roads, the Directive has been implemented in England and Wales by the Highways (Assessment of Environmental Effects) Regulations 1999 (SI No 369) http://www.legislation.hmso.gov.uk/si/si1999/19990369.htm. These Regulations replace the Highways (Assessment of Environmental Effects) Regulations 1988 (SI No 1241), except for projects begun before the new Regulations came into force on 13 March 1999.

The Regulations incorporate an exclusive threshold for the purpose of screening Annex II projects: only those in which the area of the completed works and the area occupied during the road construction or improvement period exceeds one hectare, or are wholly or partly in a sensitive area, need to be screened for EIA.

Roads developed by local planning authorities, and roads developed by private developers, require planning permission and the provisions described in Parts I and 2 of this booklet apply to such roads.

Oil and Gas Pipe-lines

Oil and gas pipe-lines with a diameter of more than 800 millimetres and a length of more than 40 kilometres fall within Annex I of the Directive. Pipe-line installations which are below either of these thresholds fall within Annex II. Oil and gas pipe-lines 10 miles long or less are approved under planning legislation.

Onshore Pipe-line Works

Proposed onshore pipes (but not those of public gas transporters-see below) which are more than 10 miles long require a pipe-line construction authorisation (PCA) from the Secretary of State for Trade and Industry under the Pipe-lines Act 1962. The Pipe-line Works (Environmental Impact Assessment) Regulations 2000 (SI No 1928), which replace the provisions relating to pipe-lines in the Electricity and Pipe-line Works (Assessment of Environmental

Effects) Regulations 1990, implement the Directive in England, Wales and Scotland in respect of pipe-lines which require a PCA. They require 'relevant pipe-line works' to be subject to EIA unless the Secretary of State makes a determination that EIA is not necessary. Public gas transporter pipe-line works

Under the licensing regime introduced by the Gas Act 1995, companies wishing to convey gas may be licensed as public gas transporters. As such, they are exempt from the need to obtain authorisation for their pipes under the Pipe-lines Act 1962.

The Public Gas Transporter Pipe-line Works (Environmental Impact Assessment) Regulations 1999 came into force on 15 July 1999 and implement the Directive in England, Wales and Scotland. They require a public gas transporter proposing to undertake pipe-line works which fall in Annex I of the Directive to submit an environmental statement and apply to the Secretary of State for Trade and Industry for consent to carry them out. Pipe-line works in Annex II of the Directive may be subject to EIA if they have a design operating pressure exceeding 7 bar gauge or either wholly or in part cross a sensitive area. In these circumstances, the public gas transporter must, before commencing construction, either obtain a determination from the Secretary of State that an environmental statement is not required, or give notice that it intends in any event to produce an environmental statement. The Regulations also provide for the Secretary of State to require an environmental statement where proposed works do not meet these criteria but nevertheless it is considered that there are likely to be significant environmental effects.

Offshore Oil and Gas Projects

The Directive covers the extraction of petroleum and natural gas for commercial purposes where the amount extracted exceeds 500 tonnes per day in the case of petroleum and 500 000 cubic metres per day in the case of gas.

For such projects, the Offshore Petroleum Production and Pipe-lines (Assessment of Environmental Effects) Regulations 1999

(SI No 360) implement the Directive. They apply to the whole of the UK. They replace the Offshore Petroleum and Pipe-lines (Assessment of Environmental Effects) Regulations 1998 (SI No 968), which are revoked except for saving provisions for applications for consent received before 14 March 1999.

The Regulations are linked to licences granted by the Secretary of State for Trade and Industry under the Petroleum Act 1998. Such licences require the Secretary of State's consent to be obtained to the drilling of a well, the getting of petroleum (where the amount exceeds 500 tonnes per day in the case of oil and 500 000 cubic metres per day in the case of gas) and the erection of any structure in connection with a development. They require, subject to an exception for floating installations whose use commenced before 30 April 1998 (when the 1998 Regulations came into force), the Secretary of State's consent for the use of a floating installation in prescribed circumstances. The Regulations also require the Secretary of State's consent to the use of a mobile installation for the purposes of the extraction of petroleum where the main purpose of such extraction is the testing of a well.

For projects which fall outside Annex I of the Directive, the Secretary of State decides whether the project is likely to have significant effects and therefore whether EIA is required.

Nuclear Power Stations

Annex 1.2 of the Directive covers nuclear power stations and other nuclear reactors, and now includes their dismantling or decommissioning, so EIA is mandatory for these projects. It does not cover research installations for the production and conversion of fissionable and fertile material, whose maximum power does not exceed 1 kilowatt continuous thermal load.

For new nuclear power stations over 50 megawatts, the Secretary of State for Trade and Industry is the competent authority, and the Directive has been implemented in England and Wales by the Electricity Works (Assessment of Environmental Effects) Regulations 2000 (SI No 1927), which also applies to thermal power stations and overhead power lines. Power stations of less than 50 megawatts, and other buildings to house small nuclear reactors, are given consent through the normal planning system and are subject to the provisions described in Parts 1 and 2 of this booklet.

The Directive's provisions on the dismantling and decommissioning of nuclear power stations and reactors have been implemented by the Nuclear Reactors (Environmental Impact Assessment for Decommissioning) Regulations 1999 (SI No 2892) which apply to England, Wales and Scotland. They prohibit dismantling or decommissioning from being carried out without the consent of the Health and Safety Executive. A licensee who applies for consent is required to provide the Executive with an environmental statement. The Regulations also apply to changes to existing dismantling or decommissioning projects which may have significant effects on the environment (Annex II.13 of the Directive).

Other Power Stations and Overhead Power Lines

Annex I of the Directive includes thermal power stations with a heat output of 300 megawatts or more, and the construction of overhead power lines with a voltage of 220 kilovolts or more and a length of more than 15 kilometres.

The construction or extension of power stations exceeding 50 megawatts, and the installation of overhead power lines, require consent from the Secretary of State for Trade and Industry under sections 36 and 37 of the Electricity Act 1989, and the requirement for EIA has been implemented as part of the procedure for applications under those provisions. The Electricity Works (Environmental Impact Assessment) Regulations 2000 (SI No 1927) require EIA for all projects which fall within Annex I of the Directive. They also require proposed power stations not covered by Annex I, and all overhead power lines of 132 kilovolts or more, to be screened for EIA. Overhead power lines below 132 kilovolts are not normally expected to have significant environmental effects and do not need to be considered for EIA unless local circumstances demand it. The Regulations also make provisions relating to nuclear power stations.

The Regulations apply to England and Wales and replace that part of the Electricity and Pipe-line Works (Assessment of Environmental Effects) Regulations 1990 dealing with power stations and overhead cables. The 1990 Regulations, however, remain in force in respect of applications lodged before the new Regulations came into force. Power stations of 50 megawatts or less are approved under planning legislation and are subject to the provisions described in Parts 1 and 2 of this booklet.

FORESTRY PROJECTS

Projects in Annex II of the Directive include `Initial afforestation and deforestation for the purposes of conversion to another type of land use.' The Environmental Impact Assessment (Forestry) (England and Wales) Regulations 1999 (SI No 2228) require anyone who proposes to carry out a forestry project that is likely to have significant effects on the environment to apply for a consent from the Forestry Commission before starting work. Those who apply for consent will be required to prepare an environmental statement.

There are four categories of forestry project that fall within the Regulations: afforestation (creating new woodlands), deforestation (conversion of woodland to another use), constructing forest roads, and quarrying material to construct forest roads.

The Regulations incorporate thresholds below which the project is taken to be unlikely to have significant effects on the environment. The thresholds are 5 hectares for afforestation and 1 hectare for other categories, although projects in sensitive areas have lower and in some cases no thresholds.

In considering whether EIA is necessary for projects which meet the appropriate threshold, the Forestry Commission takes account of the nature and scale of the project and the sensitivity of the site. Given the variability of forestry developments and the importance of location in determining whether significant effects on the environment are likely, it is not possible to formulate criteria which will provide a universal test of whether or not EIA is required.

The question must be considered on a case-by-case basis. Anyone contemplating a forestry project is advised to contact the appropriate Forestry Commission Conservancy Office for advice on whether or not EIA will be required.

An applicant who disagrees with the Commission's opinion that consent is required for a project may ask the appropriate Minister (in England, the Secretary of State for the Environment, Food and Rural Affairs) for a direction.

If the Forestry Commission discovers that work is being carried out that would have required its consent or has breached the conditions of consent, it may serve an Enforcement Notice. This can require the person carrying out the work, or someone else with sufficient interest in the land, to do one or more of the following: stop the work, apply for consent, restore the land to its condition before the work was started, carry out work to secure compliance with the conditions of consent, remove or alleviate any injury to the environment which has been caused by the work.

LAND DRAINAGE IMPROVEMENTS

Land drainage projects fall within Annex II. 1(c) of the Directive. New land drainage works, including flood defence works and defences against the sea, require planning permission and the provisions described in Parts 1 and 2 of this booklet apply to such works. Land drainage improvement works undertaken by drainage bodies are permitted development under the Town and Country Planning (General Permitted Development) Order 1995 and are therefore exempt from planning permission. As such works might have significant effects on the environment, the principles of EIA need to be applied to them. This is done through the Environmental Impact Assessment (Land Drainage Improvement Works) Regulations 1999 (SI No 1783), which apply to England and Wales and revoke and replace earlier Regulations which came into force in 1988 and were amended in 1995.

The 1999 Regulations require a drainage body to consider whether proposed improvement works are likely to have significant effects on the environment. Where the drainage body considers that there are unlikely to be such effects, it must publicise its intention to carry out the works. If the drainage body receives representations that there are likely to be significant effects but it still thinks otherwise, it must apply for a determination to the appropriate authority (Secretary of State for the Environment, Food and Rural Affairs or, in Wales, the National Assembly for Wales).

Where the drainage body concludes that the works are likely to have significant environmental effects, it must publicise its intention to prepare an environmental statement and notify specified consultation bodies (English Nature, the Countryside Agency and any other authority or organisation which might have an interest).

The environmental effects of the improvement works must be assessed in the light, in particular, of the environmental statement and any representations received. If there are no objections, the drainage body may determine that it will proceed with the works. If there are objections, the proposal must be referred to the appropriate authority for a determination giving or refusing consent to the works. The determination must be publicised.

Ports and Harbours

Annex 1.8 of the Directive includes trading ports, piers for loading and unloading connected to land and outside ports (excluding ferry piers) which can take vessels of over 1350 tonnes. Annex II.10(e) includes construction of harbours and port installations, including fishing harbours, which do not fall within Annex I.

Proposed development at ports and harbours down to the low water mark is subject to Town and Country Planning EIA Regulations. For other works, the requirements of the Directive have been implemented in England, Wales and Scotland by the Harbour Works (Environmental Impact Assessment) Regulations 1999 (SI No 3445).

These Regulations incorporate a revised Schedule 3 to the Harbour Works Act 1964, which gives the procedures for harbour revision orders and harbour empowerment orders, and replace the Harbour Works (Assessment of Environmental Effects) (No 2) Regulations 1989 governing harbour works not needing to be authorised by harbour order.

Whether or not proposed works are subject to a harbour order, the appropriate Minister (generally the Secretary of State for Transport) is responsible for deciding whether EIA is required and, if so, whether the works should proceed in the light of the assessment and the comments of environmental bodies.

In general, works under harbour empowerment orders are likely to fall within Annex I of the Directive and require EIA in every case, whereas works under harbour revision orders are likely to fall within Annex II and the need for EIA will depend on whether there are likely to be significant environmental effects. Proposed works which are subject to harbour orders are not subject to EIA if the area of works comprised in the project does not exceed 1 hectare, unless any part falls within a `sensitive area' or the Secretary of State decides that EIA is needed in a particular case.

Marine Fish Farming

Intensive fish farming is an Annex II project. Developments of onshore fish farming facilities may require planning permission and the provisions described in Parts 1 and 2 of this booklet apply to proposals for such facilities.

Off-shore facilities do not require planning permission but require a lease from the Crown Estate Commissioners (or, where appropriate, from the Shetlands Islands Council or Orkney Islands Council). The Environmental Impact Assessment (Fish Farming in Marine Waters) Regulations 1999 (SI No 367), which apply to England, Wales and Scotland, set indicative criteria which, when triggered, require the Commissioners to screen applications to determine whether EIA is needed. They apply where the development is designed to hold a biomass of 100 tonnes or more, or where it will extend to 0.1 hectare or more of the surface area of the marine waters (including any proposed structures or excavations), or where any part of the development is to be carried out in a sensitive area. The Regulations prohibit a lease from being granted before the environmental statement provided by the developer, and comments on it from consultees and the general public, have been considered.

The Crown Estate Commissioners have, in consultation with the Scottish Executive Rural Affairs Department, issued an Environmental Assessment Guidance Manual for Marine Salmon Farmers which provides guidance to developers on how to conduct EIA and prepare an environmental statement. Guidance on the practical implications of the 1999 Regulations has also been issued by the Scottish Executive Rural Affairs Department.

Marine Dredging for Minerals

Extraction of minerals by marine dredging falls within Annex II.2 (c) of the Directive. The Environmental Impact Assessment and Habitats (Extraction of Minerals by Marine Dredging) Regulations, which are likely to come into force early in 2001, will implement the Directive in waters around the UK (but excluding waters around Scotland). They will also implement the Habitats Directive (92/43/EEC).

Unless the regulator (Secretary of State in respect of English Waters or National Assembly in respect of Welsh waters) has provided, in response to a request, a written determination that proposed dredging is not likely to have significant effects, application for permission must be obtained and the application must be accompanied by an environmental statement. The only exceptions to this are where the Secretary of State has determined that the dredging would form part of a project serving national defence purposes, or where the regulator has determined that it is exempted under Article 2.3 of the Directive.

Transport and Works Act 1992

The Transport and Works Act 1992 (TWA) enables orders to be made authorising the construction or operation of railways, tramways, other guided transport systems and inland waterways; and works interfering with rights of navigation. Orders may also authorise ancillary matters such as compulsory purchase and creating or extinguishing rights over land. Applicants for TWA orders may apply at the same time for a direction that planning permission is deemed to be granted.

The procedures for making applications for orders under Part 1 of the Act are contained in the Transport and Works (Applications and Objections Procedure) (England and Wales) Rules 2000 (SI No 2190), which also implement the Directive. These Rules apply in England and Wales and replace the earlier 1992 Rules (SI No 2902) except for applications made and not determined before the new Rules come into force on 16 October 2000. In England, most applications for orders are determined by the First Secretary of State. Applications for orders relating wholly to Wales fall to be determined by the National Assembly for Wales.

The procedures include provision for opposed orders to be considered by way of a public inquiry, hearing or exchanges of written representations. Schemes that are identified by the Secretary of State or the Assembly as being of national significance are also subject to Parliamentary approval prior to a public inquiry being held.

PROJECTS ARISING IN SCOTLAND AND NORTHERN IRELAND

The arrangements for environmental impact assessment in Scotland and Northern Ireland are broadly similar to those applying in England and Wales, but are mainly subject to separate legislative provisions. The Environmental Impact Assessment (Scotland) Regulations 1999 (SSI 1999/1) implement the Directive in Scotland for projects which are subject to planning permission and for certain trunk road projects and land drainage works. Scottish Executive Development Department Circular 15/1999 explains the Regulations, and Planning Advice Note (PAN 58) on Environmental Impact Assessment also provides background information and guidance on best practice. The preceding section, of this booklet indicates which of those Regulations relating to projects outside the planning system in England and Wales, also apply in Scotland.

The Planning (Environmental Impact Assessment) Regulations (Northern Ireland) 1999 (SR 1999/73) implement the Directive in Northern Ireland for projects subject to planning permission. Development Control Advice Note 10 (Revised 1999) issued by the Planning Service, an agency of the Department of the Environment for Northern Ireland, explains the Regulations.

Appendix 8 includes a list of all implementing legislation currently in force in England and Wales, Scotland and Northern Ireland, and a list of available guidance. Go to table of contents

CONSOLIDATION OF DIRECTIVE 85/33 7/EEC ON THE ASSESSMENT OF THE EFFECTS OF CERTAIN PUBLIC AND PRIVATE PROJECTS ON THE ENVIRONMENT

As amended by Council Directive 97/11/EC (adopted by Council 3 March 1997). (Text of Directive 97/I1/EC is available in the Official Journal of the European Communities, 14 March 1997). Text in italics is a change or an addition made by 97/11/EC.

The Council of the European Union Having regard to the Treaty establishing the European Community, and in particular Article 130s (1) thereof, Having regard to the proposal from the Commission 1, Having regard to the opinion of the Economic and Social Committee 2, Having regard to the opinion of the Committee of the Regions 3, Acting in accordance with the procedure laid down in Article 189c of the Treaty 4.

1. Whereas Council Directive 85/337/EEC of 27 June 1985 on the assessment of the effects of certain public and private projects on the environment 5 aims at providing the competent authorities with relevant information to enable them to take a decision on a specific project in full knowledge of the project's likely significant impact on the environment; where as the assessment procedure is a fundamental instrument of environmental policy as defined in Article 130r of the Treaty and of the Fifth Community Programme

of policy and action in relation to the environment and sustainable development;

2. Whereas, pursuant to Article 130r (2) of the Treaty, Community policy on the environment is based on the precautionary principle and on the principle(sic) that preventive action should be taken, that environmental damage should as a priority be rectified at source and that the polluter should pay;
3. Whereas the main principles of the assessment of environmental effects should be harmonised and whereas the Member States may lay down stricter rules to protect the environment;
4. Whereas experience acquired in environmental impact assessment, as recorded in the report on the implementation of Directive 85/337/EEC, adopted by the Commission on 2 April 1993, shows that it is necessary to introduce provisions designed to clarify, supplement and improve the rules on the assessment procedure, in order to ensure that the Directive is applied in an increasingly harmonised and efficient manner;
5. Whereas projects for which an assessment is required should be subject to a requirement for development consent; whereas the assessment should be carried out before such a consent is granted;
6. Whereas it is appropriate to make additions to the list of projects which have significant effects on the environment and which must on that account as a rule be made subject to systematic assessment;
7. Whereas projects of other types may not have significant effects on the environment in every case; whereas these projects should be assessed where Member States consider they are likely to have significant effects on the environment;
8. Whereas Member States may set thresholds or criteria for the purpose of determining which such projects should be subject to assessment on the basis of the significance of their environmental effects; whereas Member States should

not be required to examine projects below those thresholds or outside those criteria on a case-by-case basis;

9. Whereas when setting such thresholds or criteria or examining projects on a case-by-case basis for the purpose of determining which projects should be subject to assessment on the basis of their significant environmental effects Member States should take account of the relevant selection criteria set out in this Directive; whereas, in accordance with the subsidiarity principle, the Member States are in the best position to apply these criteria in specific instances;
10. Whereas the existence of a location criterion referring to special protection areas designated by Member States pursuant to Council Directive79/409/ EEC of 2 May 1979 on the conservation of wild birds 6 and 92/43/EEC of 21 May 1992 on the conservation of natural habitats and of wild fauna and flora 7 does not imply necessarily that projects in those areas are to be automatically subject to an assessment under this Directive;
11. Whereas it is appropriate to introduce a procedure in order to enable the developer to obtain an opinion from the competent authorities on the content and the extent of the information to be elaborated and supplied for the assessment; whereas Member States, in the framework of this procedure, may require the developer to provide, inter alia, alternatives for the projects for which it intends to submit an application;
12. Whereas it is desirable to strengthen the provisions concerning environmental impact assessment in a transboundary context to take account of developments a tinternational level;
13. Whereas the Community signed the Convention on Environmental Impact Assessment in a Transboundary Context on 25 February 1991. (Preamble of 85/337/EEC) The Council of the European Communities

Having regard to the Treaty establishing the European Economic Community, and in particular Articles 100 and 235 thereof, Having regard to the proposal from the Commission 8, Having regard to the opinion of the European Parliament 9, Having regard to the opinion of the Economic and Social Committee 10, Whereas the 1973 11 and 1977 12 action programmes of the European Communities on the environment, as well as the 1983 13 action programme, the main outlines of which have been approved by the Council of the European Communities and there representatives of the Governments of the Member States, stress that the best environmental policy consists in preventing the creation of pollution or nuisances at source, rather than subsequently trying to counteract their effects; whereas they affirm the need to take effects on the environment into account at the earliest possible stage in all the technical planning and decision-making processes; whereas to that end, they provide for the implementation of procedures to evaluate such effects; Whereas the disparities between the laws in force in the various Member States with regard to the assessment of the environmental effects of public and private projects may create unfavourable competitive conditions and thereby directly affect the functioning of the common market; whereas, therefore, it is necessary to approximate national laws in this field pursuant to Article 100 of the Treaty; Whereas, in addition, it is necessary to achieve one of the Community's objectives in the sphere of the protection of the environment and the quality of life.

Whereas, since the Treaty has not provided the powers required for this end, recourse should be had to Article 235 of the Treaty; Whereas general principles for the assessment of environmental effects should be introduced with a view to supplementing and coordinating development consent procedures governing public and private projects likely to have a major effect on the environment.

Whereas development consent for public and private projects which are likely to have significant effects on the environment should be granted only after prior assessment of the likely significant environmental effects of these projects has been carried out; whereas this assessment must be conducted on the basis of the appropriate information supplied by the developer, which may be supplemented by the authorities and by the people who may be concerned by the project in question; Whereas the principles of the assessment of environmental effects should be harmonised, in particular with reference to the projects which should be subject to assessment, the main obligations of the developers and the content of the assessment.

Whereas projects belonging to certain types have significant effects on the environment and these projects must as a rule be subject to systematic assessment. Whereas projects of other types may not have significant effects on the environment in every case and whereas these projects should be assessed where the Member States consider that their characteristics so require; Whereas, for projects which are subject to assessment, a certain minimal amount of information must be supplied, concerning the project and its effects.

Whereas the effects of a project on the environment must be assessed in order to take account of concerns to protect human health, to contribute by means of a better environment to the quality of life, to ensure maintenance of the diversity of species and to maintain the reproductive capacity of the ecosystem as a basic resource for life.

Whereas, however, this Directive should not be applied to projects the details of which are adopted by a specific act of national legislation, since the objectives of this Directive, including that of supplying information, are achieved through the legislative process.

Whereas, furthermore, it may be appropriate in exceptional cases to exempt a specific project from the assessment procedures laid down by this Directive, subject to appropriate information being supplied to the Commission.

ENVIRONMENTAL DECISION MAKING

Public participation practice is a growing part of environmental decision-making. This study looks at its genesis in the latter half of the 20th century. It emerged from an increasing awareness of the fragility of the biosphere in which we all live, coupled with a growing desire to get involved in its protection. Emerging from the Earth Summit in 1992, Principle 10 and Agenda 21 both called for increased public participation in environmental decision-making (EDM) and led to the adoption in Europe of the Aarhus Convention, to which Ireland is a signatory. The legislation resulting from Aarhus is discussed along with its implications. The pertinent legislation is the backdrop, but many other factors affect the development and effectiveness of participatory processes. These factors along with the costs and benefits are outlined. Taking all the relevant factors into account, it is clear that effective participatory processes in EDM require good planning.

The setting of objectives, process planning, selecting the relevant public, and choosing the appropriate methods or tools to use at each stage of a process, are all part of that planning. These are all described together with methods for assessing their effectiveness. In order to see how the methods explored might be applied to a particular process, Environmental Impact Assessment (EIA) is looked at in some detail. International best practice in public participation in EIA is also examined. Methods for assessing the state of public participation on a national level are considered, and a preliminary look is taken at Ireland's status, in the light of all of these. Finally, a survey of Local Authorities, Industry, Environmental Awareness Officers and the Public looks at; attitudes to, experiences of, capacity to deliver, and capacity building measures for, public participation in EDM. The overall aim of the study was to create a picture of public participation in EDM, as practiced world wide, and to look at what is possible in an Irish context. In order to fulfill the promise of Principle 10 in Ireland, it is necessary to develop effective, creative and worthwhile public participation practice in EDM. This treatise attempts to provide a starting point for the necessary debate required to bring about that promise.

PUBLIC PARTICIPATION IN ENVIRONMENTAL DECISION MAKING

The title of this study, poses several questions: What is the 'public'? What is 'participation'? What are the 'decisions' to which it refers? Starting with the last and working backwards, EDM refers to any process of decision-making where consequent significant environmental impacts are a possibility. This includes law making, planning, strategic planning, resource management planning, licensing of industry e.g. IPPC, environmental assessment (EIA), spatial planning etc. EDM can be even more complex than decision-making on other public issues.

First, environmental impacts do not respect property, jurisdiction or boundaries.

Second, EDM can involve government agencies as both manager and regulator.

Thirdly, environmental issues can provide especially heated value conflicts that require value trade offs.

The level at which the public is involved varies with the relevant legislation, and the attitude of the other stakeholders. Often it just means informing the public of a previously made decision and asking for comments, which may or may not be heeded. Sometimes it means informed consultation. Here there is exchange of information prior to the relevant authority's reasoned decision making, and all inputs are included and seen to be included. An example of this being the River Basin Management process under Article 14 of the Water Framework Directive as described by Judith Cuff (Cuff, J. 2001). Another example would be the Bantry Bay Coastal Zone Charter. Public participation, at its apex, could also mean that the public itself, in consultation with the relevant bodies, makes the final decision itself. Examples of this can be found under the auspices of Local Agenda 21. For public participation to be effective at any level, it requires the public to be well informed and kept aware of the possibility of participation. This requires a pro-active approach from industry and the relevant public bodies.

What then is 'the public'? The public is often treated as a unitary body, whereas in reality it is a collection of numerous continually shifting interests and alliances (Ortolano), which may be in conflict with each other. The term is used as a "catch-all to describe those with an interest in a decision, other than a proponent, operator, or responsible authority". (Petts and Leach). The individuals making up a public may be involved as individuals or as members of organisations. They may become involved due to their proximity, economics, social or environmental issues, values, etc. There is further discussion on this in Section 5. By contrast, stakeholders, of which the public is one, are literally those with a stake in an issue and may include non-governmental organizations (NGO's), government or its agents, industry, individuals, communities etc.

Stakeholders do not always want to be involved in an EDM process, but they have the right to know if their interests are affected. They may want to become involved at a different stake of the EDM process.

BENEFITS, COSTS AND THE FORCES AFFECTING PUBLIC PARTICIPATION IN EDM

Public participation in EDM is playing an increasingly important role in the protection of the environment. The forces driving the increased use of participatory practice, the benefits accruing from same, as will as the costs and potential problems are described below.

Forces Driving Stakeholder Processes

Many different forces are at work, prior to, and throughout, a public participation programme. However, the following factors have also moved the participation debate forward. People generally resist change, especially when they do not understand or agree with the goals, the methods, the sponsor or the timing of the proposed change. So, keeping the public in the dark is often a recipe for disaster. Citizens are also increasingly reluctant to defer to 'expert' government agency opinions, and are unwilling to act as sounding boards for bodies that have already made decisions, particularly when they affect their local communities.

Both of these tendencies lead to a lack of public confidence and trust, unless the relevant authorities take note of them. There are growing social expectations for environmental equality resulting in part from the high media profile given to environmental issues. Findings, that minority and low income groups in the USA had been disproportionately affected by pollution and downgraded environments, resulted in the issuing of Executive Order 12898 by President Clinton.

A far-reaching programme ensued, based on meaningful involvement and fair treatment for all races, cultures and incomes in environmental decision-making. "The transmittance of 'experience', rather than knowledge, is the critical neglected dimension of decision-making, particularly as it pertains to our national decision-making"(Barkenbus, J. 1998, p.9). For example development planning in Dublin City requires largely the same expertise as rural development planning, but the experience, vision and values of the rural population may be very different to that of their city cousins. The further the EDM process is from central government, the more pertinent becomes the public participation in same, and in this context, face-to-face meetings are critical to producing real learning and trust.

Previously, with the exception of Development Plans and Waste Management Plans, public involvement in EDM in Ireland was largely restricted to the minimal fulfillment of the two "rules of national and constitutional justice, and in particular 'audi alteram partem' (hear both sides)" (Scannell, Y., Cannon, R., Clarke, M., and Doyle, O. 1999).

In practice this meant that firstly a decision was made and this was then presented to the public for comment, a process also known as 'report and comment'. So, both sides have been heard, but one has the power, and can choose to ignore the input of the other. Changes driven by the Aarhus Declaration and Principal 10 will affect many areas of environmental regulation and planning, opening them up to far greater public participation. Alongside this is the increasing transparency of regulatory processes.

The directive on Freedom of Access to Information on the Environment (Directive 90/313/EC) SI.125 of 1998 and The Freedom of Information Act, 1997 have opened up the EDM processes to closer scrutiny by the public. By compelling the relevant authorities to release documents on request, the inner workings of government have been opened up for closer examination, and relevant information freed up to enable more effective public participation in environmental governance. There is also a growing diffusion of information. The many portals of information available on the internet, together with the traditional media, libraries and the excellent work, in Ireland, of ENFO, all increase awareness of environmental issues.

Since the Aarhus Convention, which agreed to recognise and support the crucial role played by ENGO's in articulating opinions of an environmentally concerned public, European funding has been given to support their activities (Jeffrey, D.W. 2001). €4.62 million was distributed in Europe in 2003. In Ireland, the Department of the Environment and Local Government allocated €132,000 in 2001, and €133,000 in 2002 to build capacity in 23 ENGOs with a national remit.

This was done via a limited company known as Environmental (Ecological) Core Funding Ltd. (Morrisey, P. 2003). It was estimated in 1999 that there were some 500 International ENGOs (Environmental NGOs) world-wide that operated in three or more countries. Apart from the rapid growth of the ENGOs, there is clearly expanded interest in participation. Public unrest at the prospect of development, without formal opportunity for participation in decision-making, is increasingly responsible for expensive delays. In the UK, policing of antiroad protests has alone cost millions of pounds (Treweek, J.1999). The Shell Oil Company, in the UK, was eventually forced to adopt a major consultation programme on the fate of the Brent Spar, North Sea oil exploration rig, as a result of the international reaction to the campaign waged by Greenpeace. (Acland, A. 2002, 4-1).

Hampshire County Council, in England, with a population of some 2,000,000, had to scrap its waste management plan, following much public unrest, and start from scratch, with a well planned and successful public participation programme. A waste plan was then adopted which had broad support, and is being put into action. (Acland, A. 2002, 4-2). Many industry organizations are realizing that for their good image, public safety, and the longterm financial well-being of their members, responsible partnership with host communities is essential.

There is some evidence that corporations which build sound, responsive relationships with all their stakeholders tend to be longer lived and more successful as businesses in the long term than other corporations (Wheeler and Sillanpaa, 1997). Olson, M. and Toyne, P.(2000) also report a study which suggests a direct correlation between increased public involvement of the public in corporate affairs, and increased profits.

The Chemical Manufacturers Association in the U.S.A. (representing 90% of all chemical production) set up the Responsible Care Initiative, which sets codes of practice. There are six areas covered by these codes viz: Community Awareness and Emergency Response (CAER); The Pollution Code; The Process Safety Code; The Distribution Code; The Employee Health and Safety Code; and The Product Stewardship Code. In relation to all of these, the guiding principles exhort their members "to recognize and respond to community concerns about chemicals and our operations, and to work with others to solve problems created by past handling and disposal of hazardous substances." The adoption, by some industries, of environmental management systems, such as I.S.O. 14001, and E.M.A.S (Eco-Management and Audit Systems) shows their belief that a transparent and responsive industry is also more likely to be a viable one. E.M.A.S. requires public reporting of environmental management and auditing. Internationally, trends towards decentralization in many countries, increases in the numbers of countries with representative democratic forms of government, and conditions laid down by donor agencies on development loans, e.g. The World Bank, are some of the international forces bringing increased public involvement in EDM.

Whilst it is not the intention of the author to go into detail on risk assessment and disaster management., it is worth noting that public participation should, under Council Directive 96/82/EC (Seveso ?) and the Council Decision 1999/847/EC, play an important part in both of these important areas. Wates, N. (2000) gives guidelines for organizing community risk assessment and disaster management programmes, whilst the "UK Guidelines for Risk Assessment and Management" (Defra UK. 2002) give a good introductory insight into the social aspects of risk, and in particular the factors influencing risk perceptions. These factors are listed here and are also relevant to other areas of EDM.

- o Risks which are involuntarily imposed (e.g. pollution from an incinerator) tend to be seen as less acceptable than voluntary ones (e.g. bonfires)
- o Unfamiliar risks (e.g. GMO's) tend to cause concern especially when it appears that they are poorly understood by science.
- o Activities associated with a dreaded form of death (e.g. cancer) are viewed with alarm.
- o Man-made or technological risks (e.g. nuclear power) are less acceptable than natural ones (e.g. Radon). o A risk involving a single large-scale consequence (e.g. civil aviation accident) causes more concern than risks which result in numerous small scale consequences (e.g. car accidents).
- o Alarm may be caused by risks when the consequences of exposure are delayed and cause hidden or irreversible damage (e.g. ionizing radiations).
- o Inequitable distribution of risks and benefits as a result of a particular activity is likely to make the risk less acceptable. An example here might be the remaining operative Chernobyl nuclear power stations. Employees, knowing the risks involved but needing to earn a living, would have a different view of risk to others living in the vicinity, and different again to those living 100 kms away, and relying on the electricity to heat their homes in temperatures of-30°C.
- o Activities posing a threat to certain groups such as children or future generations are generally more worrying.
- o Risks which are the subject of controversy and contradictory information generally cause concern.

Benefits of Public Participation.

The protection and enhancement of the environment is the main aim of most EDM processes, and public participation can improve the quality of both the process and the end decision. The participation process is also a learning one for all the stakeholders involved, especially if there is a free flow of information between the parties.

For the public it can often be the first experience of taking an active part in the democratic process. EDM processes benefit from the direct and immediate knowledge held by citizens and business, concerning environmental conditions in their communities and industries. Encouraging the public and other stakeholders to share their knowledge, with the regulatory authorities, fosters better-informed decisions and decreases the likelihood of environmental harm, whilst increasing project viability. Of 25 overseas projects sponsored and evaluated by the World Bank, 13 failed mainly through lack of local input (E.L.I. Research Report, 1997) (Technical Document 1, 1996). It should be realized that "wisdom is not limited to scientific specialists and government officials," (120 p.6) and that "rational analysis, carried on in ignorance of political reality, may well end up so divorced form social reality, as to be of little use to anyone." (Barkenbus, J. 1998, p.2) (Acland, A. 2002).

Public participation gives broader perspectives on a particular process, and early involvement gives: added time to study issues and develop the process; enhanced credibility of the decision making process; early identification of the diverse perspectives on the issues of concern; and the generation of solution options (E.L.I. Research Report, 1997). Better-designed projects, which avoid costly delays in appraisal and implementation, can also result from early and planned consultations and public participation (Bisset, R., 2000) (Acland, A. 2002, 4-1 to 4-5). For the proponent, early participation can have the added benefit of diffusing opposition to a project. If a broad based consensus is built, it can also lead to a public sense of 'ownership' (Acland, A. 2002, 4-2). The public's enthusiasm is potentially a powerful motivating force for the EDM process. Public input can also supplement scarce government monitoring, inspection, and enforcement, resources (Bisset, R. 2000).

Involvement in EDM is a learning experience giving the public insight into the governance process. If the experience is good, it powers the way for future co-operation. The converse can also be true! Experience has also shown that environmental issues are powerful catalysts of civic 9 participation and serve as a good incentive for citizens' action and responsible democracy (Technical Document 1, 1996). Executive Order 12898 introduced by President Clinton is a far-reaching programme, based on meaningful involvement and fair treatment for all races, cultures and incomes in environmental decision-making. It has done much to prevent the dumping of dirty industry on the marginalized members of the USA citizenry. The order requires state agencies, and in particular the EPA, to pro-actively seek out the disenfranchised members of communities likely to be affected by developments, or already living in depleted environments, and encourage them to become involved in the relevant EDM processes.

Costs of Public Participation

There is limited information on the relative costs of using the different methods of participation practice, although some attempt is made to quantify these in the section 5.4. which describes each method. Actual costings for specific processes are given in a number of studies (Petts, J. and Leach, B. 2000) (Cuff, J. 2001. pps 85, 89 and 97). Clearly deliberative processes, which are used to engage relevant stakeholders in debate, discussion and deliberation, could only be done as part of a major project. Expenses range from €20-40,000 (for a citizen jury with 16 participants for 5 days), to €200-300,000 (for 3 community advisory groups of 16 people over 6 months). (Petts, J. and Leach, B. 2000, P. 41).

However, extensive use of participatory processes in the 'Wise Use of Flood Plains' project showed costs of about 10% of the above. Most of this funding was used to employ facilitators (Cuff, J. 2001). The costs to members of the public in time and money, as well as emotional stress in situations where access to EDM is limited, do not appear to have been estimated previously. An attempt is made to gather some facts about the first two in this study. The intensity of the stakeholder process, including commitments of time, energy and money, and the often, uncertain results, has led to a burnout phenomenon amongst many participants, and from all stakeholder groups. (Yosie, T.F. and Herbst, T.D. 1998) Excessive raised expectations about participation can lead to frustration and anger with a breakdown in trust. It is essential then to be clear with all participants as to the limits of influence of the process. Where an infrastructure project crosses planning jurisdiction boundaries or where more than one regulator is involved, the costs to the public, state and proponents are multiplied. The increased time scale and expense involved in public participation is of particular significance in projects with a small capital funding. An international example of this would be the AIJ projects (Activities Implemented Jointly under the Climate Change Convention). These projects already require appraisal by two governments and then there is the extra of public participations costs. After this they must compete with traditional energy supply systems, where often no regulatory costs are involved.

Problems Encountered in Public Participation Practice

The main areas of difficulty affecting public participation are attitudinal, lack of capacity to deliver programmes, lack of clarity about what outcomes are possible and the lack of a legislative framework. Some aspects of these issues are addressed below. The difference between success and failure is not always clear. If the public participation process delivers a clear decision, which is then overruled by the legislature, is this a success? (E.L.I. Research Report 1997 p. 13). A good example here is the Bantry Bay Integrated Coastal Zone Management (ICZM) Project, funded for three years under the EU Life Programme. After three years of intense community activity and the drawing up and enactment of a text-book public participation plan, the Bantry Bay Coastal Zone Charter was created.

At this point, no further funding was available, and the Project was closed down. Clearly the process was a success, but the result not. There is a general lack of experience of participatory processes in Ireland, and very few stakeholders have experienced planned participation processes. There may be suspicion, cynicism, or enthusiasm, but there is unlikely to be previous training. Suspicion about participation could be from politicians, who may feel their power is being diluted, or from ENGOs that have very specific and fixed viewpoints. Cynicism could be from previous public participation processes which were poorly managed, leaving participants reluctant to try again (E.L.I. Research Report, 1997). Citizens are frustrated when they are treated as adversaries, rather than welcome participants in the EDM process........ they feel they have been invited too late in the process. Public 'hearings' often do not include 'listening'. Citizens efforts and ideas are not included in proposals and no reason given.

The resources available to proponents and government are overwhelming. Litigation and direct action at least get a reaction" (C.E.Q. 1997. p.18). Similarly, public hearings are criticised for: being held at times inconvenient for the public, establishing an atmosphere that inhibits dialogue, and conducting proceedings that intimidate the public (Lein, K.L. 2003, p.56).

The lack of a clearly defined purpose can also means that the role of the public is sometimes unclear not only to the public but to those delivering the process, with the consequence that the participation process chosen is often unsuitable for the particular EDM process (Breggin, L and Hallman, H. 1999. p.44). The making available of the necessary resources for a worthwhile public participation process, whether by the proponent or the regulator, by the commitment of personnel and finances, is essential. The implications for project proponents, of not having a properly planned and funded participatory EDM process, may prove much more costly, than dealing with the backlash. In many circumstances, the choice increasingly is not whether to involve the public, but how to get the best value from the chosen process (Petts, J. and Leach, B. 2000).

To avoid hearing from only the activist or the powerful elite, and in order to get the widest sweep of opinion and information, authorities must reach out into the community (Canter, L.W. 1996. p.592). In this regard some activities are best avoided, for example public meetings.

In the absence of trained facilitators these can often come down to 'he who shouts loudest wins!' Lack of technical support for the public, and difficulties in getting access to information can diminish the ability of the public to play a meaningful part in EDM processes. Public access to independent expertise in a particular technical/ scientific field can prove difficult, particularly where the proponent of a project is a monopoly employer in the industry in the state. Faced with five weeks to object to a planning proposal, for example, this is a major hurdle. Added to this, it is the experience of the author that stakeholders with no technical/scientific background find it problematic to access, comprehend or evaluate data and information, especially under these short time constraints.

The feelings of powerlessness engendered by these hurdles, add to public perceptions of a lack of influence at both lower and higher levels of national and European government. "Consultations of interested parties.... can only ever supplement, and never replace, procedures and decisions of legislative bodies which possess democratic legitimacy" (COM (2002) 704. final p.4). This sense of powerlessness may be part of the reason why, even in cases where considerable energy and resources are expended, to identify individuals and groups, only a small proportion of the public ever attend participation programmes (Petts, J. and Leach, B. 2000. p.18). Public involvement may actually result in an increased level of conflict (134.p18). Becoming involved in a consultative process where the decision has already been made, or where the possible outcomes are not made clear at the outset, can lead to a great deal of frustration.

Anger may follow when the input of the public is ignored, especially following prolonged constructive engagement with an EDM process. Conflict can also occur between professionals. One source of such disagreements could evolve from the different emphases of 'scientific peer review' and 'social peer review'. The scientific peer review process is well established, and is essential to assess the technical information provided to policy makers. The parallel social peer review, designed to obtain societal acceptance and legitimacy for the decisions rendered, by contrast, does not have a set of recognised professional standards. (E.L.I. Research Report. 1997. p.14). It should also be remembered that whilst science may strive for the 'truth', the opinions of scientists are coloured by their values and beliefs. Scientists, engineers and other 'experts' should show respect in the face of 'emotional', cultural and traditional forms of wisdom. The perceptions created by a technocratic philosophy encourage NIMBY activism and further polarise communities.

The resulting deadlock frustrates the ability of planners to achieve consensus on initial issues and contributes to anti-participation attitudes. Therefore, rather than define 'NIMBY'attitudes as an irrational response to problems ordinary citizens cannot grasp, the solution is to encourage more participation not less (Lein, K.L. 2003. p.56). The misconception that the public are overly emotional, ill-informed and solely motivated by selfish interests, in their obstruction of facilities for the common good, leads to an information-based strategy. The assumption being that, 'if only the public knew the truth!' This assumes the existence of an objective truth to know (Petts, J. and Leach, B. 2000. p.9). The conflict that arises from having different bodies making decisions on different parts of the same EDM process can often confuse the public understanding and increase frustrations. For example the executive split of planning and environmental matters in some aspects of development between the Local Authority and the EPA. Public involvement in these circumstances must deal with two different agencies (Canter, L.W. 1996. p.592) (Gray, A.J. et al 2002. p.6).

Organisations resist change, and communication blocks within organisations can prevent them from making timely responses to their publics. The established routines and organizational systems of many bodies seek to promote the 'status quo'. Organisational structures which support sustainability, and therefore public participation, are a threat to 'command and control' style management systems. Managers, who have risen to power in this style of structure, frequently resist the transition to alternative structures, which embrace public consultation (Griffiths, A. 2000). Many opportunities for public participation are laid down in the environmental legal framework, as shown in Appendix 1. Where legislation does not specify early public involvement, then it is at the discretion of the relevant authority, and dependent on the opinion of the relevant public servant.

For example, in the development planning process, the Manager's reporting of the consultative process clearly includes the names and nature of all submissions together with his response to them (Section 11 of the Planning and Development Act 2000), whereas in making an Air Quality Management Plan (under Section 48 of the Air Pollution Act, 1987 and Section 102 of the Environmental Protection Agency Act 1982) there is no legal compulsion to show how any submission might have affected the final decision.The recent use by Cork City Council of Article 81 of The Planning and Development Act, Regulations, 2001, to enter into a public/private partnership with a company previously refused planning permission to develop the same site (TP 02/26234), leaves the public only minimal scope for comment and no route to appeal any decision made.Under section 19 of the Protection Of The Environment Act, 2003, the making, variation or replacement of a Waste Management Plan, has been made an Executive Function. This has removed the immediate political accountability of the decision-making, and so removed another route for public input.

GENERAL HISTORICAL PERSPECTIVE

General provisions for public access to information, participation and justice go back 14 centuries in Islamic Law, as recognized in many countries of North Africa, The Middle East and Asia. (Ahmad, A., 2002) but the International Boundary Waters Treaty, 1909, was the first global treaty to include strong provisions for public access to information and participation (Bruch, c. and Filbey, M., 2002). Broad international recognition of public involvement including provisions for transparency, participation and justice dates back 55 years. The Universal Declaration of Human Rights in 1948 provided the kernels for generalized access to information (Article 19) and Justice (Articles 8 and 10) as well as the right to associate (Article 20). In the West widespread public awareness of changes in the balance between human life and the environment emerged in the1960's, with the publication of Silent Spring (Carson, R. 1962), The Population Bomb (Ehrlick, P.R.,1968) and Small is Beautiful (Schumacher, E.F., 1973).

The immediacy of television portrayals of such disasters as the Torrey Canyon and "acid rain" (Jeffrey, D.W., 2001) made an even greater impact. In the USA as in much of the West, the 1960's and 1970's were marked by a participation movement (Breggin, L. and Hallman, H., 1999) which led to legislative changes. The Freedom of Information Act, 1966 gave access to data and information held by government agencies. The National Environmental Policy Act, (NEPA) 1969 amongst its measures required public review of Environmental Impact Statements (EIS). Previously the Administrative Procedures Act of 1946, directed government agencies to introduce a notice and comment stage into rule making procedures and to keep registers.

This Act also allowed for the right of appeal to federal courts (Ortolano, L., 1997) So together these three pieces of legislation provided the foundation for the involvement of the public in environmental protection in the USA, and NEPA has largely been the basis on which environmental impact assessment world-wide has been based. The voice of environmental concern was internationalized at the Stockholm Conference, 1972 (United National Conference on the Human Environment), drawing together both developed and developing nations, but in the absence of the Soviet Unions and most of its allies. For the first time environmental non-government organizations (ENGO's) played a important part in the deliberations.

Stockholm was the event that turned the environment into a major issue at the international level, declaring the right of people to live "in an environment of a quality that permits a life of dignity and well being". (GEO-3. UNEP., 2002) Amongst its impacts were:

- Some 50 governments together with the Organisation for African Unity (OAU) have recognized the environment as a fundamental right.

- During 1971-1975, 31 major national environment laws were passed in OECD countries, compared with 32 in the previous 15 years.
- By 1982 the number of countries with environment ministries rose from 10 in 1972 to 100.

Although the link between environment and development was made in Stockholm, it was the 1983 World Commission on Environment and Development (WCED) also known as the Bruntland Commission that first developed the idea of Sustainable Development, in its broadest sense. The WCED report in 1987 resulted from consultations at all levels of the global society, and attempted to balance the needs of future generations with those of our own. In essence it tried to promote both the inter and intra generational equity (GeO-3 UNEP., 2002). However, in its concluding statement, the commission stated that "existing decision making structures could not cope with the demands of sustainable development" (WCED.1987). The UN Conference on Environment and Development (UNCED or "The Earth Summit") held in Rio De Janeiro, Brazil, saw 176 government, 100 heads of state, 10,000 delegates, 1400 NGO's and 9,000 journalists gather in what is still the largest event of its kind ever (GEO-3.UNEP., 2002). The summit was preceded by global consultations with hundreds of thousands of groups and individuals.

Amongst its many achievements, the two most relevant here are, Agenda 21, and Principle 10 (P.10), of the 'Rio Declaration on Environment and Development'. Agenda 21, in its 40 chapters, lays a solid foundation for the promotion of sustainable development. Chapter 8, Section 1, calls on governments to seek broader public participation in policy making and decision making for sustainable development. Section 3 is devoted to the strengthening of the roles played by the principal social groups: women, children and young people, indigenous populations; NGOs; local government; workers associations and unions; business and industry; and the scientific and technological community (NCEDR, 2003) (Agenda 21, Chapter 8.).

Chapter 40 establishes two programme areas to ensure that decision making is based on sound information: namely, bridging the gap in availability, quality, coherence, standardization and accessibility that exists between developed and developing countries; and improving availability of information. Principle 10 states:

"Environmental issues are best handled with the participation of all concerned citizens, at the relevant level. At the national level, each individual shall have appropriate access to information concerning the environment that is held by public authorities, including information on hazardous materials and activities in their communities, and the opportunity to participate in decision-making processes. States shall facilitate and encourage public awareness and participation by making information widely available. Effective access to judicial and administrative proceedings, including redress and remedy, shall be provided".

It would be difficult to overstate the significance of Principal 10 in providing a mandate to regional and international governance development in the decade following the Rio Declaration, as well as to an underlying commitment motivating national law making. Since 1992 a raft of regional and international treaties have incorporated P.10, whilst addressing such issues as Desertification, Bio-Diversity and Clean Development Mechanisms.

International organizations such as the World Bank, and to a much lesser degree the World Trade Organisation, have made steps towards making public consultation and involvement an integral constituent of their development loans. In the context of this study and from an Irish and European perspective, the most important initiative evolving from P.10 is the 1998 UN/ECE Convention on Access to Information, Public Participation in Decision-Making and Access to Justice in Environmental matters. The ECE (Economic Commission for Europe) is a regional body of the United Nations and includes 55 countries. The European Community plus 39 other nations have signed the convention (also known as the Aarhus Convention), which it was ratified in 2001. The Aarhus Convention is based on the three pillars, ensconced in P.10 and elaborated in

Agenda 21. The pillars are; access to information; public participation in decision-making; and access to justice.

The Aarhus Convention

The Aarhus Convention sets out to put a shape on the aspirations of Principle 10 (P.10). It encapsulates the three main pillars of P.10, access to information, public participation in decision-making, and access to justice, and sets the basic tenets for all European legislation on EDM. Pillar I-Access to Information. Article 4 of Aarhus sets out the general right of persons to gain access to existing information, on request. Pillar II-Public Participation in Decision-Making. Article 6 covers public participation (PP) in decisions on specific activities, e.g. on the proposed locating, construction and operation of large facilities. Article 7 deals with PP in the development of plans programmes and policies relating to the environment, which includes sectional or land use plans, environmental action plans, and environmental policies at all levels. Article 8, covers public involvement in the preparation by public authorities of laws and regulations.

In each of the Articles, 6, 7 and 8, early involvement of the public is encouraged. Articles 7 and 8 are less precise than Article 6. Article 6 provides for a high level of involvement adequately guaranteed by law (Stec, S. and Casey-Lefkowitz, S., 2000. p.86). Article 6 can apply to spatial planning decisions, development consents, operating permits, discharge permits, or any particular proposed activity where the decision making may have a potentially significant impact on the environment. Article 6 is not limited then, to activities where an EIA is required or to those activities listed in Annex 1 of the Convention.

Pillar III-Access to Justice

Article 9 allows the public to challenge "acts and omissions" by either public authorities or private persons, as applied to articles 4 and 6, as well as allowing them a review of violations of domestic environmental laws. In relation to Article 6, members of the public must have a significant interest {Article 9 (Section 2(a)} or maintain impairment of a right (Article 9 section 2 (b)). A level of standing is given to both individuals and NGO's. The provisions of an expeditious, free or very cheap, alternative review mechanism established by the courts is to be made available. Publicity about these procedures and removal of barriers to access are also stressed. It is clear that in order for Pillar II to operate, Pillars I and III must also be well established, i.e. access to decision making cannot be effective without good information resources and the possibility of enforcement through access to justice. It should be noted that Article 3 reminds the parties that the provisions of the convention are only minimum requirements.

The Implementation of Aarhus

The implications for environmental governance under European environmental law and consequently that of the Republic of Ireland are far reaching as will be seen below. Pillar I-Access to Information. National Level. The Freedom of Information Act, 1997, (FOIA) gives access to all information held by the state, including information on the environment, other than the exceptions listed in the Act. ENGOs, however have told the author that they see S.I.125 of 1998 as a more effective tool. Directive 90/313/EC (on public access to information on the environment), implemented in Ireland as S.I. 125, of 1998, (The Access To Information On The Environment, Regulations, 1998.) has already been repealed by the European Commission and replaced by Directive 2003/4/EC, which must be brought into Irish law by 14th February 2005. The resultant changes include:

- Persons or bodies which perform functions by arrangement with a public authority and which effect the environment are themselves deemed to be public authorities for the purposes of the Directive (Article2, Section 2(c)), and so should make available environmental information held by or for them on request.
- Applicants for information do not have to state an interest.
- Information should be made available in the format requested (Article 3, Section 4).

- "Course of Justice" exceptions refer to information that may adversely affect the course of justice (Article 4, section2 (c)) rather than the previous "sub-judice" exceptions.
- Definite response times and appeals procedures are laid down. EC Institute Level.

Regulation (EC) 1049/2001 grants European citizens and residents a right of access to documents of the European Parliament, Council and Commissions. Applicants receive an acknowledgement of receipt, and within 15 days provide the item(s) requested or give reasons why not. The applicant is entitled to complain to the Ombudsman or institute Court proceedings if a negative reply is received. 20 pages of A4 text will be photocopied free of charge and any extra charged at cost.

Pillar II. Access to Participation

National Level

The proposal for a Directive of the European Parliament and Council providing for Public Participation in respect of drawing up of Certain Plans and Programmes relating to the Environment, and Amending Directives 85/337/EC as implemented in Ireland by S.I. 349of 1989 (and amended by Directive 97/11/EC), and 96/61/EC. These two refer to "Environmental Impact Assessment" and "Integrated Pollution Prevention and Control", respectively. The new directive will make changes in PP procedures required by these directives as well as creating a requirement to consult the public in drawing up of plans or programmes in seven other directives viz:

- The 1975 Framework Waste Directive (75/442)
- The 1991 Batteries Directive (91.157)
- The 1991 Agricultural Nitrate Directive (91/676)
- The 1991 Hazardous Waste Directive
- The 1994 Packaging Directive (94/62)
- The 1996 Ambient Air Quality Directive (96/62)
- The 1999 Waste Landfill Directive (99/31)

Honouring the subsidiarity principle, the proposed directive sets out the minimum requirement for effective public participation, in fulfillment of the objectives laid down by the EC Treaty in the field of the environment, leaving the detailed arrangements to the member states. The proposed directive requires member states to take the necessary measures, to ensure that the public are given early and effective opportunities to participate in the preparation (and review) of the plans and programmes listed above. For this purpose, "the public" has the same meaning as in Article 2(4) of the Aarhus Convention. To that end, Member States are required to ensure that:

- o the public are informed (by public notices or other appropriate means) about any proposals for such for such plans or programmes and that relevant information about such proposals is made available to the public.
- o the public are entitled to express comments and opinions before decisions on the plans and programmes are made;
- o and that in making those decisions, due account is required to be taken of the results of the public participation.

Taking the EIA directive 85/337/EC as an example, the following changes are proposed. The Aarhus Convention definitions of "the public", and of the "public concerned" are inserted into 85/337/EC. The public concerned will be defined as the public affected by or likely to be affected by, or having an interest in, the development consent procedure. A new Article 6(2) requires Member States to take steps to ensure early and effective opportunities to participate. The new Article 6(3) will lay down requirements for informing the public based on Article 6(2) of the convention. This requires the disclosure of all pertinent documents and opinions expressed by relevant public authorities.

Reasonable time frames for the stages of participation should also be provided. Changes to Article 7 of the EIA Directive regarding Transboundary EIA are discussed. The IPPC Directive will have amendments similar to those of the EIA Directive, with the addition of a new Annex V in which detailed provisions for public participation will be set out. Both the Water Framework Directive (2000/60/EC) and the proposed Directive 2001/42 on "the assessment of the affects of certain plans and programmes on the environment (SEA Proposal), already include provisions which are consistent with Article 7 of the Aarhus Convention (Banfield, N., 2003.). EC Institute Level Although no legislative proposal exists at present, COM 2002/704 Final, entitled, "Towards a reinforced culture of consultation and dialogue-General principles and minimum standards for consultation of interested parties by the Commission", provides good insight into the governance practice being developed within EC structures.

By fulfilling its duty to consult (Protocol No.7 annexed to the Amsterdam Treaty), "the Commission will help to improve the quality of policy outcome and at the same time, enhance the involvement of interested parties and the public at large". (COM.2002.p.5). The commission has underlined its intention to "reduce the risk of the policy makers just listening to one side of the argument or of particular groups getting privileged access" (COM.2002 p.5). These minimum standards laid down in the communication were themselves the result of a European wide consultative process.

Pillar III Access to Justice

National Level

Both Directive 2003/4/EC and the proposed Directive 2000/0331(COD) amending 85/337/EC and 96/61/EC, contain provisions for access to justice in line with the Aarhus Convention. Changes to the other directives listed are given in Annex 1 of the proposed directive 2000/0331(COD). However, shortcomings in the enforcement of environmental law throughout Europe, as outlined in the Sixth Environment Action Programme (6EAP), (Com (2002) 31, Final), will be the subject of a new Directive being worked on by the commission at present (Banfield, N.,2002). All of these measures will become part of Irish Law in due course, when transposed by the Dail. In 6EAP it is recognized that better access to courts for NGO's and individuals would have a number of helpful effects in relation the implementation of community law. EC Institution Level.

3

THE CONTRIBUTION OF ECONOMIC POLICY

INTRODUCTION

The Ecologically Sustainable Development Working Group on Forest Use specified three main requirements for sustainable forest use:

- maintaining the ecological processes with in forests (the formation of soil, energy flows, and the carbon, nutrient and water cycles);
- maintaining the biological diversity of forests;
- and optimising the benefits to the community from all uses of forests within ecological constraints. The National Forest Policy Statement adopts these principles as the basis for ecologically sustainable development. (National Forest Policy Statement (NFPS), 1992:47) (Emphasis added.)

Economic theory sets a standard by which the outcomes are measured, and informs us as to what are 'good' and 'bad' ways of making choices. Economic theory also provides an important underlying philosophy which specifies that changes should be to the net benefit of society at large.

THE ROLE OF ECONOMIC POLICY

Economic policy is often seen as a method, or science, of analysing impacts on the financial e c o n o m y. When properly and rigorously deployed, however, it is elevated to the 'art ' of choice.

Thorough economic analysis concerns itself with how to make choices that provide the greatest net benefit to the community, where 'benefit' implies consideration of all social values. It is for this reason that economic theory, incorporating such concepts as externalities and public goods, has come to play a significant role in decision making processes. The following sections briefly outline the relationship between the concepts of externalities, economic value and economic decision making.

EXTERNALITIES AND THE WELFARE OF SOCIETY

Economics is centred on the allocation of scarce resources to competing ends, in order to achieve the most value for human society. In this instance, economic analysis seeks to identify the flows of value that can be derived from 'uses' and 'non uses' of the forest.

This process of analysis seeks the combination of uses and non uses that most meets the value preferences of society, both now and in the future. The process is intended to be value neutral in the sense that it is based on the given or expected values of society-whatever those values may be.

Traditionally, economic theory uses the 'market' as its benchmark of efficient resource allocation; that is, an allocation of resources that creates the greatest benefit for society. This 'pure' model, however, admits that its results hold true only under very tight and specific conditions. If any of these conditions are breached, then there can be no presumption that the free market outcome is the best one.

Externalities arise when certain conditions of a free market do not hold. They arise when the free market fails to include, or take account of goods and services provided incidentally by the completion of another transaction. An externality can be compared to an auction where the main transaction is completed between two parties, but where additional goods and services, not provided for in the original transaction, are subsequently delivered.

These can be delivered either to the main parties or to third parties who were not involved in the original deal, and who, indeed, may even be unaware of the original transaction. For example, if a tree is cut down in the forest for timber, the loss of that tree may adversely affect the view of a home owner, who subsequently experiences a loss in his/her property value.

Externalities and public goods are both expressions of market failure. 'Public goods' refers to a value from whose benefit others cannot be excluded. Thus, deriving pleasure from knowing of the existence of a forest is an externality, because others cannot be excluded from deriving the same pleasure. At the same time, the phenomenon is also a public good because it is non rival in consumption; that is, many can consume the item without reducing the consumption of others.

In order to avoid confusion, all such values are called 'non financial values' since they cannot, by definition, be traded in a market place, and hence be expressed in terms of financial value. Where environmental issues are concerned, the market cannot be relied upon to produce an optimal allocation of resources-ie. the greatest overall benefit to the community-because there is no venue, no 'global auction room', in which the community (in particular future generations) can express its preferences over the items it values.

The 'items of value' in this discussion are the 'values' or 'benefits' that flow from the 'use' of 'objects'. Normally, economic analysis talks of objects such as goods and services when the real interest is in the values that flow from these objects. It makes little sense to talk, for instance, of a 'forest' per se. Are we referring to the wood alone, the biological species that reside there, the pleasure that bushwalkers derive, or its function in absorbing greenhouse gases? The term clearly has no operational meaning unless we refer explicitly to its various use values and benefits. In order to avoid any ambiguity, the remainder of this paper will generally use the term 'values' rather than 'goods' or 'products' (as 'values' is a more inclusive term).

An understanding of externalities is central to quality economic decision making, particularly in the context of the environment-simply because in the real world, there are many items of value (fresh air, knowing that an old growth forest exists, etc) that cannot be bought and sold in the market. The exclusion of these items from the 'economic equation' means that the overall equation is miscalculated. Miscalculations of the total economic equation generally lead to a misallocation of economic resources, and to an overall loss to society.

The practical significance of identifying externalities and public commodities relates to their impact on markets. It is impossible for a market alone to systematically produce the socially optimal outcome when externalities and public commodities are present, as they are in the case of the natural environment.

Pricing Theory and the Financial Values of the Forest

In economic theory, the price of an item can represent one of three circumstances for the owner.

- The price could reflect the value of the item to the owner.
- The price could value the item at a level greater than the owner's subjective valuation. In this case the owner is likely to sell the item.
- Alternatively, the price of the item may be less than the value to the owner in which case the owner will refuse to sell.

The price of wood must also reflect the different type of values generated by the wood and the portion of forest which is altered by its extraction.

Hence, if the 'public citizen' 2 offers a logging coupe for sale, by definition the value anticipated from the sale must at least exceed the value expected from any alternative use. And additionally, the price needs to equal:

- the costs of the production of wood;
- plus a return on the capital invested, with sufficient reward to cover the risk; and
- total compensation for all non financial values (NFV's) lost in the process of an extraction of the timber.

This pricing method ensures that timber with values exceeding the market price will not be sold. However, the significance of NFV's is that they make up a considerable proportion of the value of timber. When NFV's are destroyed, the loss of value is not felt by one person alone, but is likely to be felt by every Australian citizen, every international citizen and everyone yet to be born. When logging occurs, each of these people experiences a change in the value they derive from the forest, and hence each should be compensated. The price of timber needs to be able to compensate all losers if policy is to be socially optimal.

The pricing of timber is crucial to determining the pattern of logging in the forest. In order to implement a solution, the present approach of forestry commissions uses regulation to separate logging areas from non logging areas. This approach could be assuming one of two things. Firstly, it could be assuming that royalties are sufficient compensation for the loss of NFV's 3. In this case, it remains to be demonstrated by those who set royalty levels that they have evaluated the worth of the lost NFV's, and thus set royalty levels accordingly. In the absence of information about how NFV's are determined, it could be assumed that non financial values in the logged areas are considered insignificant to the extent that the Australian community will not require compensation for the losses in NFV's which result. In theory, logging will then occur, in the areas set aside, if the cost of production and capital makes it a financially rational exercise.

Pricing policy is, thus, conducted on a 'compartmentalised basis'. In other words the forest is surveyed for its values, and those areas where NFV's are deemed 'small enough' to avoid the need for compensation, are made available for logging. The remainder is put aside for conservation. Thus, if we use a 'compartmentalised' approach, timber is valued in a two stage process. The forest is allocated to different uses depending on the scale of NFV's in different areas. Forest that is to be subject to logging is then priced on a commercial basis. Pricing takes place on the assumption that the coupes have already been surveyed, and the NFV's have been found to be minor; ie very close to zero.

The worth of the royalties, of course, must also cover the costs of wood production, the return on capital required, and the financial opportunity cost of foregone uses. The Victorian Auditor General's report (1993) indicated the likelihood that the forestry operations in Victoria were making low profits (if any at all). Given that royalties must cover operating, capital and opportunity costs, this leaves very little room to compensate for lost NFV's. In such a circumstance, it is feasible that NFV's are being valued at a worth very close to zero. No evidence was identified to justify such a valuation.

The analyst must be careful to ensure that the BCA has the proper scope. Stand-level analysis is appropriate for landowners who are interested in single stand management and even for larger ownerships such as conservation authorities which do not require an annual timber harvest. Furthermore, stand level results can be usefully incorporated into forest level harvest scheduling programs when a relatively small proportion of the management unit area is affected by the treatment alternatives. However, if a manager is comparing treatments that will be applied to a significant number of hectares in a forest, then a forest level analysis is needed to properly account for between stand interactions. This type of analysis can be very complex and envolved and usually requires the use of computer modelling.

The intensification of a regeneration regime should yield a merchantable stand sooner than would otherwise be the case and presumably will contain a higher yield of a more valuable species. The benefits then will be the increased value of the future stand due to the higher stocking to the desired species and the earlier avialability of the areas for harvest.

Typically, the point at which a stand-level analysis begins is after the removal of the current stand. The rationale for this is that the money spent to regenerate the site (i.e. the investment) is related to objectives concerning the future stand, and it is on those merits that the regeneration expense should be evaluated. When spending more money on the harvest operation will reduce the costs of subsequent regeneration, the extra expense should be included as a regeneration cost, since this is the reason for it.

In a single stand analysis, one should ensure that there are no inconsistencies in the analysis caused by evaluating different alternatives over different time periods, as can happen when different actions will affect the stand rotation age. This can be avoided by examining a long series of repetitions of each alternative, effectively giving rise to an analysis period of infinite length. Alternatively, one could choose an analysis period based upon multiples of the harvest ages (e.g. compare two fifty-year harvest cycles under intensive management versus a one hundred year harvest cycle under extensive management).

However, it is not always simple to find the required multiples, especially when a large number of options are being compared. This type of analysis is accomplished by considering the present net worth of a series of harvest values predicted infinetely into the future.

The value of wildlife is more difficult to assess. Forest management will likely shift the distribution of wildlife species in a stand or forest. Ideally, one would like to have estimates of how wildlife populations will be affected by alternative management actions. This is difficult, due to the number of species involved and the variations that populations exhibit over time. It is not yet possible to track all species-only those of importance or value to humans are tracked (although the criteria for importance are changing rapidly).

At the present, valuation of a major species poses significant obstacles. Suppose the moose population is locally important. Moose have value because people like to view them and because people like to hunt them. In theory, it is possible to survey people who live in and/or visit the area of interest, ask what value they place on having various population levels of moose, and adjust the results to account for the entire local (human) population and tourists. Statistical refinements based on income and level of camping or hunting experience are often made.

Studies in the U.S. have shown that a typical person will be willing to pay some amount for higher populations of many species because there are better wildlife viewing opportunities and because there is some satisfaction in knowing that there are more (rather than fewer) animals of that species alive in the area. A hunter will also be willing to pay some amount to have more game species in an area because the chances of a successful hunt are increased. A larger game population will also be likely to lead to an influx of hunters from other areas, which is a plus in a regional analysis but of little effect in a provincial or national analysis.

The author is not aware of any published data based on willingness-to-pay surveys that have estimated the value of wildlife species in Ontario. Where game species are involved, one alternative is to use the value of expenditures made by hunters as a proxy for the hunting value per animal. This has the obvious drawback of ignoring the value of benefits unrelated to hunting, but since expenditure data are available, they represent the best alternative for now. Of course, the benefit-cost analysis must still relate population level to degree of hunting effort and this will require some estimation.

The valuation of recreation opportunities poses similar problems. However, the key steps are the same as in the case of wildlife and timber. First, a quantitative relationship must be established between alternative management actions and recreation opportunities. Secondly, the value of a recreation opportunity, usually measured in terms of a recreation-day, must be estimated. One estimation technique is the willingness-to-pay survey. A more common technique, which is being employed by the MNR Forest Values Group, is known as the travel cost method. The central idea is to identify where the majority of recreation users come from, divide this region into zones of known distance to the recreation site, estimate the cost of travelling to the site from each zone, and then use the incidence of recreational use per zone and the zonal population to construct a demand curve for recreation at the site. Using site entrance fees is not appropriate for valuation purposes because prices are often set at low levels to encourage use. Finally, the valuation of such features as aesthetics is in its infancy, especially in Canada. Willingness-to-pay surveys of various types are again used in this case. Unfortunately, there are no known credible estimates of the value of aesthetic quality in Ontario.

There are two general classes of costs: out-of-pocket expenses and indirect losses or disbenefits. The out-of-pocket costs of undertaking various activities are usually unambiguous, but can be difficult to track down. Some judgement has to be exercised in the case of expenditures for fire protection and road construction and access. In these cases, a clear understanding of the reasons for making these expenditures is an important starting point. Fire protection is intended primarily to prevent the loss of human life, and secondly to prevent the destruction of buildings, machinery, or other capital goods. Therefore, only a fraction of fire protection costs can be charged against forest management.

At a forest level the choice of the exact fraction of the fire program's costs to allocate to forest management would depend upon the fire district, proximity to habitation and recreation values, wildlife concerns and of course the value of the forest and its contribution in providing wood or forest land supplies. Any attempt at a local division of these cost allocations would be artifical due to way these costs are accounted for at the aggregate program level, the interdependency of the social values associated with the use categories and the fact that in most areas fire prevention or control expensess are incurred to protect several use categories at one time.

Similarly, roads are built to facilitate timber extraction, subsequent assessment and regeneration, and provide access to the public for other purposes. Again, a subjective estimate of the proportion of the road value attributable to timber extraction is usually made. The costs of roads must also be amortised over some period of time, such as the 20 to 25 years often used for large capital expenditures such as pulp and paper mills.

As was mentioned above, costs of damage or reductions in non-timber values should also be included. These can be estimated using the principles that are appropriate for estimating non-timber benefits, described above.

The valuation of costs and benefits will depend on who is doing the analysis. A planner working for a Forest Management Agreement (FMA) holder will have a different perspective than a government planner on the benefits and costs associated with alternative management strategies. For example, since some regeneration costs are partially refunded by the government on FMAs, the company and the government will naturally place a different cost on regeneration. This is as it should be. A simple example will be presented which illustrates the use of benefit-cost analysis to compare the application of three forest-level strategies on a management unit. Suppose that these strategies are characterised as:

- Reliance on Natural Regeneration
- Extensive Management
- Intensive Management

The particulars of the strategies are unimportant; the key point is that each increase in management intensity results in a greater volume of wood being produced from the unit. Suppose that the wood supply levels are expressed as annual volumes. The table below shows, for each option, the annual wood supply, the value of the annual harvest, and the annual cost of management. Note that wood has been valued at $10/m3, and non-timber impacts have been excluded.

It will be assumed that some harvesting will be done, and so one of the three options must be selected. The key issue is the comparison of the net gains from management intensification. The table below shows the PNW and BCR associated with each option.

A comparison of PNW's shows that the greatest net benefit is associated with intensive management. Therefore, if sufficient funding is available, intensive management would be preferred. However, the BCR declines as management is intensified from extensive to intensive levels. This indicates that the gain per dollar spent decreases as the management level is increased.

If there are other potential investments that yield a BCR ranging from 1.46 to 1.60, then the greatest net benefit will be obtained by adopting extensive management and investing the remaining available funds in these other projects.

In this way, the BCR acts to guide investment so that the greatest overall PNW can be obtained. In this example, the profitability of management was at least partially justified by the increase of current harvest made in the expectation of future yield increases due to management. This is the allowable cut effect (ACE). Economists observe that the ACE has often been used to justify poortreatment investments (see Davis and Johnson, 1987, Pearse, 1990). It is considered that there are really two separate decisions involved-the selection of the harvest level and the selection of the appropriate silvicultural treatment. Mainstream economists argue that these decisions should be analyzed separately on their own merits, whereas forest economists point to sustainable yield requirements as the basis for linking the two decisions (Hyde, 1980).

In Ontario, the argument is clouded further by the public ownership of forests and the flow of non-timber benefits that come from them. This implies that the government should include in the analysis the social benefits that can be expected to arise from an increased harvest level.

What conclusions can be drawn from this debate? The first is that one should be reluctant to include the ACE when analyzing projects on private land where the sole purpose of investing is to grow timber but ACE is a valid factor to include in analyses of public forest management. The second point is to heed the economists' warnings and critically examine the merits of the silvicultural projects being considered. Foresters should be careful about planting on medium and poor site classes, and on sites distant from the mill. Due to the additional costs of hauling it to the mill, the value of timber grown further from the mill is lower than the value of equivalent timber close to the mill.

One of the great difficulties in forest planning is dealing with the long time periods required to grow timber. The fastest growing boreal stands may be harvestable in 40 to 50 years and it is difficult to forecast real timber prices that far in the future with any accuracy. The uncertainty inherent in predictions for 100 years or more is enormous. Therefore, in view of the significant degree of uncertainty associated with assessing the profitability of a regeneration project, one may question whether such an exercise is worthwhile.

Nevertheless, most people believe that planning is useful and planners can only use their best estimates. It is also the case that accurate forecasts of future prices and costs are less important when one is using BCA to select the most attractive project or to allocate funds among projects.

In these situations, any errors in forecasting will apply equally to similar projects and reservations about long-term forecasting are less relevant.

Forestry debates present society with several choices that need evaluation. The task is to estimate the value in each of the choices presented. The purpose of this paper is to establish a framework that will allow each environmental policy option to be systematically and comprehensively assessed for its impact on the total economic welfare of the community.

Such a framework consists of ensuring that the non financial values, financial values and alternative financial values derived from the forest are accurately evaluated, and hence given their full and due consideration.

There are presently significant barriers to the achievement of this objective. The three barriers identified in this report are:

1. The existing inability to determine the correct financial values for forestry operations.
2. An inability to determine the correct financial worth of alternative activities such as plantations and water production.
3. The lack of a methodology which determines the relative worth of non financial values (NFV's) of the forest resource.

Removal of these impediments, which is an exercise in microeconomic reform, will bring improvements in the financial efficiency of the Australian economy, whilst improving the overall quality of life for Australians in general. The first two impediments can be removed by the utilisation of some fairly common methods. With specific reference to the third impediment, however, there are many approaches by which appraisal of non financial values can occur.

Some of these are specific methodologies such as contingent valuation and public ranking of options. Other methods centre around public involvement in the assessment of options. Yet other methods are based on using 'rules of thumb' to account for non financial values.

Among the reports surveyed, some techniques are mentioned quite frequently. These included public participation as a key necessity for getting non financial values properly assessed. Surveying of public opinion, systematic reviews of natural features and the use of experts also seemed to be relatively common.

The task therefore becomes one of finding an appropriate route through the range of techniques offered.

Section 3 of this report briefly outlines the relationship between the concepts of externalities, economic value and community well being. The absolute requirement of good economic analysis is the need to accurately appraise all values, including externalities (non financial values) in any assessment framework. Economics provides the theoretical standard by which the existing assessment frameworks can be judged.

Section 4 reviews several major forest reports in order to identify four significant components of forest project appraisal. These include the values of the forest, the valuation methodologies proposed for assessing values, the pricing/ financial valuation of forestry and alternative activities, and the assessment framework for appraising policy options in the search for the socially best outcome.

In particular, Section 4 examines the role of pricing in the appraisal of forest policy options. The existing assessment processes and the various forest reports (apparently on the grounds of practicality) adopt a 'compartmentalised' approach to pricing. In other words, protection of non financial values is carried out prior to, and separate from, the pricing of forest timber. If the appraisal process is incomplete, then the pricing of timber is inadequate by definition. Pricing policy is significant for estimating value to the extent that it partly determines the financial value of forest timber.

Section 5 then describes weaknesses in the existing appraisal processes. These flaws range from a broad neglect of non financial values in project assessment to more specific issues. These include the failure to comprehensively identify and value non financial values, often even in the form of physical impacts; failure to identify or consider alternative financial values and alternative project options; and the failure to provide accurate statements about the financial viability of projects.

This report makes clear that there is indeed evidence of serious weaknesses in the determination of the financial values of native forest timber. Our analysis indicates the very strong likelihood that native forest timber harvesting, as a commercial activity, is being subsidised (eg. low cost public sector loans, no commercial charges for land), is inefficient (eg.. there are better returns to be had from harvesting water in at least one case) and poorly managed (eg. the agencies are unable to provide appropriate accounting information on their commercial activities).

Consequently, a significant microeconomic reform process needs to be undertaken, if the correct financial values for native timber sales are to be determined.

Inappropriate appraisal of financial values may also be suppressing economic growth in the private forestry and agroforestry sectors (and other alternative/substitute financial values). The lack of a 'level playing field', with respect to the cost of business inputs (eg. land and capital) between the public and private forest managers, suppresses investment in private forestry at a lower level than that which could normally be expected.

If financial values and alternative financial values of the forest were to be appropriately appraised, this may identify alternative revenue streams and investment opportunities for hard pressed forestry commissions.

Summarised, inaccurate financial values and a limited knowledge of alternative financial values combine to undermine the basis of comparison between the financial and non financial values of the forest. This problem is compounded by the lack of a methodology for determining the relative worth of non financial values.

In conclusion, it must be reiterated that any assessment of the social benefits and costs in forest policy must include three pieces of vital data: an accurate appraisal of financial values, alternative financial values and non financial values. Without reference to such data, environmental decisions become less an attribute of good management, than of reliance on sheer blind luck.

Resolving these problems will contribute to better environmental management, greater financial performance and more jobs. This makes reform in this area a key part of the process of microeconomic change, resulting in better environmental and economic policy.

THE THEORY OF TOTAL ECONOMIC VALUE

In any public decision over the net benefits of forests, externalities and public commodities, along with project and alternative financial values, need to be taken into account, as part of the value provided by a 'resource'. In analysing different resource options, a total listing and weighting of values, both private and 'externalised', must be provided for each option.

When an economist is comparing alternative social arrangements, the proper procedure is to compare the total social product yielded by these different arrangements... In devising and choosing between social arrangements we should have regard to the total effect. This, above all, is the change in approach which I am advocating.

Coase refers to 'total social product' which we have renamed 'total economic value' in the tradition established by Pearce (1989). In order to compare the total economic value of differing policy options, it is necessary to aggregate for each option all the values (financial, aesthetic, spiritual, etc.) that would be produced for the community if a particular policy option were to be undertaken.

Traditional project evaluation normally compares one option against another option, and chooses that which has the most value. In the presence of market imperfections such as externalities, the total economic value (the project's total costs and benefits to society) can be severely miscalculated in at least two ways. First, proper economic evaluation needs to consider the financial impact that one project may have on other parties, not simply whether it is profitable when considered in isolation. Suppose, for example, that the government is requested to give its approval for the establishment of a factory on a river bank. Viewed in isolation, the proposal is attractive: it is expected to provide jobs for 100 local unemployed young people and will contribute an extra $5 million to tax revenues. Viewed from a wider perspective, however, the project's net financial value would be seen to be negative if the poisonous effluent it emits cause the bankruptcy of downstream fish farms and tourist resorts which together generate 200 jobs and $10 million in tax revenues. These 'down stream' financial values are often overlooked by traditional project evaluation methodologies.

The second weakness of the traditional evaluation approach is that it gives little consideration to non financial values. To continue with the illustrative example above, the factory's pollution may poison wildlife and household pets, ruin a picnic spot favoured by thousands of local residents, etc.

These types of costs are real and relevant, notwithstanding the fact that they are of a 'non financial' nature. Resolving the first problem is easier, since it may be possible to obtain the relevant financial data if appropriate economic modelling is conducted. The second problem is more difficult because non financial values are not traded in markets. In other words, there is no 'objective dollar price' which can provide a yardstick with which to compare non financial and financial values. Environmental economists have been working on providing techniques to evaluate non financial values.

Economic Decision Making

As a society, and as individuals, we are constantly evaluating and rejecting options. A social cost benefit analysis is, in effect, the act of making choices. Social cost benefit analysis becomes a conceptual framework that expresses the less formal processes we use to make choices. While the National Forest Policy Statement leaves open the practical question of assessment, it identifies the need for evaluation of policy options. In Diagram 1 we have listed all the values, in the tradition of Pearce (1989) and Young (1992) that appear to be provided by the native forest. Table 1 describes each type of value, and where relevant, their relationship to each other. Total economic value (TEV) analysis would require that each policy option would receive a 'score' in each cell of an evaluation matrix (see Table 5 for an example). Scores are aggregated by a means yet to be determined which then enables decisions over differing options to be made.

Economic decision making seeks to choose between the differing options on the basis of determining that which provides the greatest net social benefit. The means of deciding which option has the greatest net social benefit where 'scores' cannot be aggregated into a single numeraire is, of course, highly contentious.

There are three channels through which resources can be allocated. These are the markets, institutional processes and executive decision making. Markets, however, as decision making institutions, are largely unavailable in the circumstances covered in this report because we are dealing specifically with instances of market failure. However, modified markets (ie. markets subject to government intervention) are included as a means of policy implementation. The choice between government modified markets and 'pure' market measures should be justified on the grounds of net social benefits.

The irony is that if market measures are to be chosen as the decision making process, then it is a decision that will have to be made by government anyway, and hence does not relieve the public decision making process of the need to evaluate the optimality of value flows from the forests under differing policy options. In other words, choosing to 'do nothing' is as much a decision as choosing to 'do something'. The lesson is that if optimal social benefits are sought, then market failure automatically triggers government decision making processes-even if that results in a decision to 'do nothing'.

The creation of a formalised process for considering environmental values and other externalities requires a clear understanding of the integrated structure of externalities, economic value and economic decision making. All values-financial, ethical, ecological, etc-must be accounted for, if different policy options are to be compared on a equal basis. The different values that comprise each policy option, need to be integrated in some manner which enables socially optimal choices to be made.

APPRAISING THE VALUES OF THE FOREST: THE THEORY AS PRACTISED

This section reviews thirteen recent reports on forests for their views on the appraisal of forest values. Each report was reviewed in order to gain an understanding of the value assessment problems in the forest sector. The reports were examined on the basis of four criteria:

- Identification of externalities.
- Identification of valuation methodologies.
- Identification of pricing/financial valuation practices.
- Identification of likely assessment frame works.

Problems in the Identification of Externalities

A representative example, of the values identified, is listed below. These values were derived from the Resource Assessment Commission (RAC) study. This study, amongst those examined, was the most comprehensive and thorough in its treatment of the issues. The weaknesses in its study were repeated, and magnified, in other studies.

The Resource Assessment Commission Inquiry into Forest and Timber

Conservation; wilderness; recreation; Financial values such as wood, tourism and water supply, oils, seeds, tannins, medicinal plants, sandalwood, bark products, fishing, hunting, wild foods, infrastructure corridors; Information via education, research; intrinsic values like aesthetic (different from visual), heritage, spiritual, biocentric, wilderness, lifestyle; Utilitarian values; 'pristinity'; landscape; scenic quality, public sensitivity, industry value by implication; existence values, option values, vicarious use values, bequest values, ecological sustainability, ethical values.

The main weaknesses in assessing the manner in which values had been surveyed were that:

- Values were not identified or listed in a systematic or comprehensive manner. The above extract from the RAC report was amongst the most comprehensive, but even so, lacked a systematic listing of values.
- Different value categories often overlapped. For example, ecological sustainability was repeatedly confused with 'environmental amenity'.

- There was also repeated confusion over the values we impute to an object, and the values we derive from that object as a result of our preferences. For instance, a forest site may have great scientific value, but not be of great aesthetic value. Conversely, a forest site may have limited scientific value, but great aesthetic value. In the latter case, a purely scientific view point would not reveal the aesthetic importance of the site. Similarly, there are other values-for example existence value-held subjectively by the public with respect to native forests. A study which evaluated objective characteristics only would not pick up the subjectively felt values. We have identified this as the 'object subject' tension.

Identifying Suggested Valuation Techniques

Each report suggested, implicitly or explicitly, a range of valuation techniques for providing an 'objective' assessment of the values identified. Overall, the lack of comprehensive identification of values, common to all thirteen reports, subsequently inhibited the matching of valuation techniques to appropriate categories of value.

This complicated the process of value assessment since it provided limited guidance about how various values should be measured.

On the positive side however, the Resource Assessment Commission (RAC, 1992) report did provide an assessment of each evaluation technique, which marks a very useful first step towards providing a listing of techniques relevant to the assessment of each value.

The Organisation for Economic Co-operation and Development references (OECD, 1986; 1992) also made a worthwhile contribution towards evaluating techniques.

In general, the reports provided only limited advice on the means by which different types of values should be evaluated. Moreover, with the exception of the RAC and the OECD reports, the techniques put forward were generally insufficient to measure the full range of values derived from the forest.

For example, the reports would discuss techniques for measuring the scientific values of the forest, with out discussing other measures that evaluate the 'worth' of the non-scientific values.

A second issue worthy of further discussion in each report was the role of uncertainty in complicating the assessment of values.

Australian & Overseas Experience of Assessment Frameworks & Processes

There are several assessment frameworks used in Australia for forest decision-making. Most of the reports have only considered assessment frameworks in general terms, without specific reference to real examples. In order to evaluate these assessment frameworks, the report provides a 'checklist'.

The 'checklist' provides a comprehensive series of questions which any worthwhile assessment framework should be able to answer. The 'checklist' approach facilitates identification of the broad strengths and weaknesses of existing or proposed assessment frameworks

A common theme in all reports (both Australian and international) was an underlying philosophical assumption that a 'technical', non-political choice process was possible, and to some extent desirable. This process was not challenged explicitly in any report, despite the obvious claim that the political arena is the desired place for settling value disputes (indeed it could be argued that this is the raison d'etre of politics). For the moment, however, this discussion accepts the premise that a 'technical' or 'black box' approach is feasible for value choices.

So as to gain an understanding of international experience, the report has sought to review processes suggested by the OECD and the World Bank. These internationally-recommended approaches provide a benchmark for achieving 'world best practice' in this field. The report has also examined the type of system deployed by New Zealand under the auspices of the Resource Management Act (NZ, 1991).

The international reports reflect a worldwide concern over the assessment of environmental values. While it is now recognised that environmental values must be incorporated, it is not known how this should be achieved. The World Bank view is not unusual:

Externalities are thus clearly troublesome, and there is no altogether satisfactory way to deal with them. This is no reason simply to ignore them, however; an attempt should always be made to identify them and, if they appear significant, to measure them. (World Bank, 1992:23)

This accurate, but unhelpful commentary is largely due to a neglect of environmental factors within economic analysis. This was noted by two internationally renowned economists in 1988, with respect to the field of development economics:

Environmental resources appear in this literature about as frequently as rain falls on the Sahara. (Dasgupta and Maler, 1990:102)

They also noted, as recently as 1988, that one of the world's best known texts on development economics (see note below)

... simply has no discussion of environmental resources and their possible bearing on the development processes.' (Das gupta and Maler, 1990:102)

This is symptomatic of neglect within the broad field of economic analysis. The OECD (1992) remarked that conventional cost benefit analysis would not include a number of environmental effects ...because of the lack of market price data or easily accessible direct proxy price data.' (OECD, 1992:49)

Given this neglect the OECD report (1986) on the public management of forests, must be regarded as a landmark publication amongst the reports reviewed in this study.

Overall, given that all economic activity centres on the idea of 'development', the general lack of incorporation of environmental issues is a serious omission. As a result the weaknesses identified in Australian assessment processes are not unusual when compared with 'world best practice'-though this does not make such deficiencies acceptable according to economic theory. Neo-classical economic theory would state quite clearly that on a worldwide basis, as well as locally, such an omission must be regarded as a threat to the economic well-being of society.

Existing Assessment Frameworks

Australia, with its federal political structure, has many different means by which choices over values are made. Rather than describe this diverse array of assessment frameworks, this section will provide a thumbnail sketch of the key features of assessment frameworks and processes. The review of various reports conducted in the previous section, along with the international examples cited, seem to point to three key characteristics of concern in any assessment:

Who Makes the Choices?

'Who makes the decisions' is significant because it determines whose values are brought to the process. The economics profession has long argued that institutions tend to be captured by the values of specific groups, and that to expect any public institution to be value-free and working purely in the public interest is naive. From the perspective of assessment processes, it is important that advocates of differing values are given an equal voice in the process of analysis and decision-making.

What Value Information do they have?

The RAC Forest and Timber Inquiry report took some interest in the Victorian model which involved the Land Conservation Council in decisions about selection of forest logging coupes. The AHC-CALM model of joint survey work also received some favourable attention, though its joint deliberations were designed to identify values rather than make decisions. That no other systems were singled out for favourable comment, in the various reports reviewed in this study, seems to indicate that other processes may be lacking an adequate contribution from representatives of other values.

It is not unfeasible that an appropriately structured institutional process could make optimising decisions-given that the right information on values was available. The information provided, as outlined by the thirteen reports reviewed in this process, is inadequate for the task of making choices over values. The AHC-CALM model received some favourable comments for its process from the ESD and RAC reports Certainly, its assessment method is systematic and thorough. Its prime weakness is that it does not consider, in its own words, 'social values' (AHC, 1992:29). 'Social values' as defined by the AHC, appear to cover that group of values described by economists as existence value. As a result, it only acknowledges some of the values considered important in a total economic assessment. That it was commended as a model for others suggests that even analysis of 'non-social' values is deficient in existing assessment frameworks and processes.

Other reports such as the Victorian Auditor-General's report (AG, 1993) and the NSW Public Accounts Committee (PAC, 1990) confirm the impression that only poor data is available on values.

Financial values are also a significant part of the analysis. A significant range of reports located deficiencies or made recommendations that pointed to opportunities to improve the gathering of financial data. The deficiencies in this regard extended to the data on both financial values generated by forestry and financial values generated by other activities. The existing assessment framework and processes were not being supplied with the relevant data because the necessary mechanisms such as appropriate accounting systems are not adequate (eg. RAC, 1992:429).

How do they Compare Different Values?

Assessment frameworks must make choices between values. Few technical models were offered on how such choices should be made,outside the political process. The Victorian Auditor-General's report noted that the previous system in that State had used a committee of relevant officials, backed up by a requirement to find options, with a final 'safety valve' being provided by there being recourse to the Minister. No doubt similar methods are used elsewhere, though, what is required is a method that considers all values at the point of decision with appropriate systemic checks and balances.

The AHC-CALM model and the process offered by the RAC Forest and Timber inquiry were the closest Australian examples of technical approaches which are beginning to seek full consideration of all values. The OECD (1986) report is a benchmark in this field. Only the process implied in the RAC report comes close to this analysis. It is assumed that if there were another system, the RAC inquiry would have located it. The existing assessment frameworks, obviously, can be said to rely on the judgements of those who take the decisions and the information that they have been supplied with. There must be serious doubts that these existing frameworks have been properly structured. A assessment framework and process needs to specify who decides, with what value information and by what intervalue comparative techniques.

Weaknesses in the Existing Assessment Framework

Inadequacies in Pricing/financial Valuation

The significance of non-financial values to the calculation of financial values is considerable. The forest agencies and all the reports reviewed in this study have adopted a compartmentalised approach to pricing. The compartmentalised approach assumes that assessment of choices between values occurred prior to the pricing of timber. If non-financial values are improperly assessed, then the resulting effect is that financial values will also be inappropriately founded. In this sense a compartmentalised approach to pricing is a practical manner of settling commercial pricing issues. In simple terms, if non-financial values are being compromised it is due to the lack of adequate assessment prior to the commercial pricing process.

Pricing of logs should, however, be offered on a normal commercial basis. It is fairly clear from the reports that have been reviewed that operations are not presently proceeding on a commercial basis. Prices of forest timber are being distorted by a range of factors that are, in general terms providing a subsidy to the logging of native timber, whilst other factors are raising the costs of private plantations (see NPAC, 1991). Other factors such as the lack of rational markets for the pricing of water (see Read and Sturgess, 1992) and other services, partly or wholly derived from forests, are raising the relative cost of alternative financial values.

The following distortions in commercial pricing of timber have been identified. The list has been 'cobbled' together from all the reports reviewed in this survey.

1. The lack of marginal cost pricing.
2. The non-existence of effective commercial accounting systems.
3. Long-distance transport subsidies.
4. Non-payment of rental for use of public land.
5. Low royalties; ie. low log prices for timber (see note below)
6. Possible cross-subsidisation from community service obligations.
7. No liability for payment of a resource rental tax.
8. No liability for payment of a notional income tax (presumably meaning company tax).
9. No liability for local government rates.
10. No liability for sales tax on vehicles, plant and equipment.
11. Non, or low, payments of interest on borrowed capital.
12. Non-payment of dividends.
13. Low user charges for cost recovery from long-distance transport
14. The lack of commercial discipline due to lack of a corporatised commercial structure;
15. Serious deficiencies in the tax system for plantations;
16. Problems with planning, rating and exporting from plantations;
17. Problems with the small numbers of buyers and sellers of timber;
18. Problems with the purchaser behaviour of large buyers of timber;
19. The dominant market role of state forestry agencies;

20. Failure to properly account for non-financial values in pre-logging operations either by exclusion of areas or by forestry codes of practice.
21. Failure to adequately assess the potential financial contribution from sales of goods and services such as water from the native forest.
22. Failure to separate commercial operations from regulatory functions.

They also felt that the market price was too low and that international prices were not as significant in pricing native forest timber as was considered by the Commission. (PAC, 1990:77) They felt that the native forest timber price should be raised to 'the point of indifference.' (?)(PAC, 1990:77) and that the case for doing so was 'overwhelming' (PAC, 1990,79). This is even more significant when it is considered that 'the Forestry Commission sets these prices...the Commission is clearly a price leader.' (PAC, 1990:97)

Inadequacies in Assessment Frameworks

The existing assessment framework is in a period of transition. The post-1945 charter of the forest agencies to push for economic development has changed. The capital resources of government and the Australian economy are scarce and need to be deployed for maximum financial benefit or for some clearly identified social 'good'. Changes in social values are also placing pressure on the old system. These two forces, in combination, are leading to greater scrutiny of forest activities, and demanding greater levels of accountability. It is clear that assessment processes are lagging in their response to these issues.

Entrepreneurial Capture

The existing assessment process is vulnerable to 'capture' by entrepreneurs proposing one-o ff developments. This limits government ability to consider a range of project options as proposed by the OECD (1986) and the World Bank (1992) in particular. The existing Australian approach of having a single project option placed in front of decision-makers must be regarded as inferior, if not a structural distortion of the decision-making process.

Lack of Commercially based Accounting of Projects

The lack of commercially-based accounting data for project analysis means that it is not possible to accurately evaluate true project financial worth. This means that many unprofitable activities are slipping through the policy net adding to the economic burden already being borne by Australians. The effect of such methods is to inflate the relative worth of financial values vis a vis non-financial values. It also inflates the relative values of forestry operations vis a vis the alternative activities such as plantations and water production.

Lack of a Comprehensive set of Values Descriptions

The next major flaw in assessment processes is the lack of a complete set of values relating to forestry. Each report examined has mentioned a range of words that seek to describe, in particular, the non-financial values, but there has not been a report which systematically listed what society derives from the forest. It is little wonder then that assessment frameworks barely exist or function if the values are not even identified.

Lack of Focus on the Full Suite of Values

Where non-financial values are considered, they tend to relate only to the assessment of physical characteristics of the forests. All reports acknowledge the need to protect these physical characteristics, for example 'sustainable yield', yet there is evidence that they are not being protected, even at this rudimentary level ((PAC, 1990:55). If physical characteristics are only barely being protected, then it is reasonable to assume that other more abstract and intangible values of the forest are not being taken into account. The Dorrigo Management Area Environmental Impact Statement (FCNSW, 1992:i) does not mention any non-financial values other than soil, birds, animals and recreation. Ecological sustainability and existence values are barely considered at all.

Public Involvement: Preached by Some but not Practised

The practice of assessment seems to place too little stress on the need to obtain public review and input into forestry decisions. In an area where clashes over values are common, the need for representation by different values should be considered. As a result the political process is starved of information about social preferences. The OECD (1986) reveals a considerable commitment to public involvement, in particular their example of the 'Inner Valley Road Project'(OECD, 1986:34).

NFV's are not being Assessed

The most serious flaw is that it appears that assessment of non-financial values is simply not occurring. The concentration of reports on discussing aspects of how such analysis could be done, rather than a focus on how it is being done leads to the conclusion that serious analysis of the relative worth of non-financial values is yet to begin. It would seem fair comment that if relatively simple choices between wood and water production are not being evaluated (eg. Read and Sturgess, 1992), then other more complex interactions are also being neglected.

Future Generations Considered by Implication Only

The OECD (1986) report was not without its flaws. Its assessment framework gave little attention to considering the impact of future generations from any valuation assessment process. In economic theory it is impossible to internalise the interests of future generations simply by assessing the preferences of the present generation. Neither market processes or valuation techniques, in the face of irreversibility, are capable of 'looking after' the interest of future generations in any manner that can be relied upon. Fortunately the concept of ecological sustainability goes some way to achieving this goal. The internalisation of the interests of future generations requires explicit discussion.

Most reports steered clear of identifying procedures that would resolve the choices thrown up in the forestry debate. The RAC report (1992) went furthest in this regard but tended to educate by demonstration rather than explicit prescription. The ESD report (1991) had some brief comments, but if the assessment of these different values is to proceed satisfactorily, it is clear that the hard issue of 'assessment frameworks' must be dealt with honestly.

The OECD (1986 and 1992) indicated that other nations are moving towards direct evaluations of non-financial values. Australia's approach would appear to be somewhat slower than average, though of course there are some institutions such as the World Bank that have acknowledged the need for change but are yet to make the practical changes that are required. It was significant that the United States was acknowledged as having greater experience in the area of non-monetary evaluation. The experience was deemed to be due to the 1969 Environmental Policy Act which required quantification in non-monetary terms of all environmental impacts. The OECD concluded that it would seem that education and more valuation projects are a prerequisite for more effective utilisation of the techniques. A learning process is needed. (OECD, 1992:38).

Valuation techniques, whether monetary or non-monetary, form the backbone of an assessment framework. The same learning process is required for assessment processes as well. Many OECD countries are learning but Australia does not appear to have started.

Assessing the Assessment Frameworks

A major difficulty with evaluating an assessment framework is the sheer scale of the task itself, as anyone who has read an Environmental Impact Statement will testify. The result is a tendency to focus on one or two key angles to the exclusion of other angles. Table 6 seeks to provide a systematic 'framework' for assessing assessment frameworks. A series of questions is provided, that an assessment framework, if it is to be successful, should be able to address. The existing Australian and international assessment practices are evaluated The table has been tested on actual reports, in this case the Victorian Auditor General's report (AG, 1993) as well as an analysis of the Dorrigo Management Area Environmental Impact Statement (FCNSW, 1992).

The native forest is capable of producing three major categories of values. These are:

- financial values from timber production;
- alternative financial values from other forms of forest use; and

- non-financial values.

  This report has sought to examine the issue of giving each type of value 'full and due consideration' in policy processes. Three themes have recurred throughout this study.

- There is a theme focusing on the analysis of technical comparisons between different values. That is the use and abuse of various techniques that quantify values.

The native forest is capable of producing three major categories of values. These are:

- financial values from timber production;
- alternative financial values from other forms of forest use; and
- non-financial values.

  This report has sought to examine the issue of giving each type of value 'full and due consideration' in policy processes. Three themes have recurred throughout this study.

- There is a theme focusing on the analysis of technical comparisons between different values. That is the use and abuse of various techniques that quantify values.
- Asecond theme revolves around the nature of the institutional arrangements that ensure value selection which is socially optimal.
- The major theme, however, is achieving a level playing field between the major types of values derived from the forest.

The emphasis on achieving a level playing field is in the spirit of the microeconomic reform process that has taken place over recent years, and continues with recent reports such as the Hilmer Inquiry (NCP, 1993). Sims described microeconomic reform as occurring when changes are made to achieve more output from a given level of inputs. (Galligan et al, 1993:15).

The output from the forests is intended to maximise the social welfare of the community. For this reason, microeconomic reform in the forest sector must seek to arrange the three sets of values above so that welfare is maximised.

The implication of the need for micro-economic reform is that the sector is not making an optimal contribution to the quality of Australian life. This review of reports on the forest sector suggest that there is considerable potential for microeconomic reform in two dimensions.

- In broad social terms it would appear that the community is not maximising the social benefit from the forest resource.
- Secondly, within the social benefit the community receives, the financial value to be had from the forest resource is not being maximised, assuming that all other types of values remain unchanged (ceteris paribus 'All other things being equal.').

If the microeconomic reform process (optimising social benefits) in the forest sector is to succeed, the estimation of the worth of the three categories of value described above must be accurate. It would appear that estimation of these three categories of value is presently inaccurate. This inaccuracy would confound any attempt to derive optimal social benefits.

Effects of Miscalculating Financial Values

The impact of miscalculating the financial value of timber production is significant for Australia's macro-economic performance. Firstly, it overestimates the financial worth of timber production vis a vis some other form of commercial activity. The Read Sturgess report (1992) made it clear that it was more financially valuable for the Upper Thomson catchment to supply water than to supply timber, and consequently more profitable for the economy. The overestimation of the worth of forestry would tend to hide this potential economic 'free lunch'. Other competing uses for the forest may be similarly affected. More generally, the over-valuation of forestry financial values disguises what may be net losses in this sector which are actually diminishing the wealth of Australia unlike other commercially viable industries.

Failure to accurately value forestry also leaves the commercial operations of the forestry commissions vulnerable to an income shortfall in the event of corporatisation. Even without corporatisation, the poor measurement processes for forestry blind the commissions to other profit centres such as leasing forests to supply water to regional communities with expanding industries.

Over-valuation of forestry through poor input pricing, in addition to affecting competing uses of the forest, also undermines industries that provide substitutes for forest timber ranging from plantations to steel production. Inefficient input pricing in native timber production undermines all competing industries, all other things being equal. This report has focused on the 'collateral damage' caused to the plantation industry. However, what applies to the private plantation industry, also applies, probably on a lesser scale, to other industries that compete with native forest timber. The private plantation industry (NPAC, 1991) is one clearly visible victim of an 'unlevel' playing field in the forest sector.

The over-valuation of native forest harvesting undervalues, by definition, the product of plantations. This leads to lower levels of investment, employment and wealth creation than may otherwise have occurred. It prevents the development of value-added timber industries that can further develop on the basis of a plantation resource. This presents the Australian economy with a 'double loss'. The macro-economy loses, marginally at least, from harvesting native forests, and then loses again through the disincentive effects on plantations.

Other industries may suffer as well. Finally, there is a third loss built into the playing field, which are the microeconomic distortions, such as poor taxation arrangements, identified by NPAC (1991), that inhibit the plantation industry. The microeconomic distortions in the plantation sector alter the financial calculus such that the value of native timber is enhanced relative to plantation timber. In addition these microeconomic distortions cause a further round of macro-economic losses from the plantation industry.

The structural distortions within the forestry sector, and within other sectors both competing and substitute, compound the overvaluation of forest timber relative to other values. This causes a loss of income, jobs and a waste of capital as private agents respond to these inappropriate signals.

The Effects of Miscalculating NFV's

The second major impact of overvaluing forestry financial values, is felt on non-financial values. Over-estimating of forestry financial values also underestimates, by definition, the relative worth of non-financial values. Thus the process of evaluating non-financial values is seriously handicapped from the beginning. The significance of this was pointed out by Sims, in another context, when he stated

Clearly our objectives are broader than maximising GDP. For example pollution in our cities is of general concern. It might be that including externalities in prices would decrease environmental damage and decrease growth, but improve overall living standards. (Galligan et al, 1993:17, emphasis added)

Two points flow from the comments by Sims.

Firstly, Sims appears to be giving emphasis to a commonly-held view that environmental improvement can only be achieved at a cost to the economy. This review has indicated, in a very clear manner, that in the forest sector financial gains (see note below) can be achieved whilst making environmental improvements. The view espoused by Sims presumes that the 'balance' between environmental and financial considerations is 'efficient'; that is, in the context of this report, the evaluation of the financial worth of native forests has been accurately calculated. The evidence indicates that the balance is inefficient, because calculations of financial and alternative financial values, as well as NFV's are inaccurate. Therefore, this report indicates that improvement in environmental conditions, over a very large margin, will improve financial conditions as well.

At some point along the policy continuum, improving environmental conditions will negatively affect financial conditions- but the forest sector, along with many other sectors, is not at this point, and indeed is a long way from it. Microeconomic reform in the forest sector has the potential to bring environmental improvement in its wake by removing financial inefficiencies that have contributed to over-exploitation of the environment.

The implication of financial 'inefficiency' in the sector needs to be clearly understood. It implies that society could be financially better off if certain microeconomic reforms were undertaken. The implication of 'economic' inefficiency is that, in the context of this report, the environment and the financial wealth of society, could be better off, leading to an unambiguous improvement in society's level of 'satisfaction'. In this sense, when 'financial inefficiency' is present, microeconomic reform and environmental policy can be complementary, and not competing policies.

Secondly, failure to evaluate non-financial values relative to financial values accurately is critically important to maximising 'overall living standards'. The benefits that are provided to Australians from NFV's can also be described as 'psychic income' (see note below).

These benefits also have the fortuitous 'collateral' benefit of also attracting tourists, and hence contributing to the expansion of this critical export industry.

Non-financial values provide direct inputs into the daily consumption of Australian citizens and even to foreigners through tourism, wildlife films and existence values. An assessment framework which seeks to optimise benefits to the community, but which does not recognise these non-financial values can generate financial flows that actually undermine community well-being.

This happens because the source of much 'psychic income' (ie. the natural environment) has been lost and replaced by financial benefits that are too small to compensate. If non-financial values are to be given their full and due consideration, then financial and alternative financial values must be correctly evaluated.

This review has identified that there are significant inefficiencies in the estimation of value in the forest sector. The presence of these inefficiencies in the market activities of the forest sector is due to structural distortions.

The structural distortions contribute to poor financial and environmental performance. There is a considerable social, environmental and financial dividend to be achieved by microeconomic reform in the forest sector. Microeconomic reform requires that a 'trilogy' of assessments be made: evaluate financial values (FV's), evaluate alternative financial values (AFV's) and evaluate non-financial values (NFV 's). Continuation of present practices such as failure to 'estimate value' accurately will cost Australia in lost jobs, lower incomes and a diminished environmental quality of life.

Here this report distinguishes between financial gains that contribute to the overall health of the financial economy, and economic gains that contribute to the overall state of social well-being. Improvements in financial well-being are commonly referred to as economic gains, for example improved resource allocation, because of the easy but methodologically sloppy habit of referring to the financial activities of the Australian society as the 'economy'. Accurately the economy is the process by which we gain 'welfare', a part of which is the financial gains offered by financial activity.

In another circumstance, when a Treasury officer was asked to described the non-pecuniary benefits of working in Treasury, he called these benefits 'psychic income'.

THE ECONOMIC DIMENSION OF TRANSITION

One way in which an economist familiar with cost-benefit analysis (CBA) may look at environmental impact assessment (EIA) is that EIA is a CBA requir-ing a large amount of information rigorously collect-ed and analyzed by a group of scientists of various disciplines. One can appreciate that the development of EIA has served to advance the systematic descrip-tion and quantification of environmental effects in a way which can only serve to improve the quality of CBA undertaken.

One can also appreciate that EIA constitutes a process, formalized through environ-mental legislation, regulations and procedures, which in turn reinforces the community participation re-quired for informed decision-makers seeking the common good. While CBA within EIA is a key gen-erator of information to the decision-makers, recognition should also be given that the EIA process is both a planning tool and an execution monitoring tool. Thus, the role of CBA is not as an intermediate step, but rather as an integrated parallel process.

Hence, the CBA should follow up, quantify and eval-uate the dynamic iterations created by the exhaustive search of alternatives, by the efficient inclusion of en-vironmental protection and mitigation measures and by the explicit valuation of preference and choice articulated by community, society and their representa-tives acting in the best interest of present and future generations.

VALUE OF CONSIDERING ENVIRONMENTAL EFFECTS EARLY ON IN THE PROJECT CYCLE

The main purpose of the economic analysis of a project is to ascertain whether the project can be ex-pected to create more net benefits than any other, mutually exclusive option, including the option of not doing it. Consideration of alternative options, therefore, is a key feature in proper project analysis. Often, important choices about alternative project options are made early on in the project cycle.

These options may differ considerably in their general economic contribution, and they may also differ greatly concerning their environmental impact. Therefore, including environmental effects in the early econom-ic analyses, however approximately, should improve the quality of future decisionmaking.

The CBA process-defining objectives, searching for alternatives, costing out the resources involved, speci-fying the effects of each option involved and compar-ing all the costs and benefits-normally requires considerable efforts. In the case of Cuba, we are faced with two joint families of economic valuation prob-lems: the "environmental" ones and those problems associated with the incipient transition from a cen-trally planned economy to a truly participative mar-ket based economic system.

In principle, economic analyses are to take into ac-count all costs and benefits of a project. With regard to environmental impacts, however, there have been two basic problems even in developed countries.

First, environmental impacts are often difficult to measure in physical terms.

Second, even when im-pacts can be measured in physical terms, valuation in monetary terms can be difficult. In spite of such difficulties, a greater effort needs to be made everywhere to 'internalizes' as many environmental costs and benefits as possible by measuring them in money terms and integrating these values in the economic appraisal.

The measurement in money terms are made even more difficult in countries undergoing market reforms.

The environmental valuation problems can be sum-marized as follows:

VALUE OF ENVIRONMENTAL ASSETS

While man-made and human capital may be valued with relative ease by observing existing market sys-tems, where available; the existence value of clean wa-ter and air, tropical forests, wetlands, coral reefs and other environmental assets and their functions is much more difficult since not even market prices can reflect their full contribution to other economic ac-tivity and to human welfare.

In particular, the mar-ket price of water does not reflect the various services it provides nor do market values can accurately re-flect what happens when irreversible loss or damage of natural resources occurs as environmental degrada-tion exceeds a critical threshold level.

Complexity of environmental valuation also arises due to the multi-ple functions of being a source of raw materials and energy, being a sink for assimilating man-made wastes and providing other services such as recre-ational/tourism services, storage of genetic diversity and scientific and educational benefits.

The following classification is useful:

- Direct use values are derived from the economic uses made of the natural system's resources and services. Examples of these are outputs such as timber, game and recreation from forests or fish and scuba tourism from coral reefs.
- Indirect use values are the indirect support and protection provided to economic activities and property by the resource system's natural func-tions or environmental services. Examples of these are watershed protection and soil erosion

prevention provided by forests, and beach sand and mooring facilities protection provided by coral reefs.

- Non-use values lie in the special attributes of the natural system as a whole, its cultural and heri-tage uniqueness; it includes both existence and option values. Existence values reflect public goods which can be enjoyed by more than one consumer without decreasing the amounts en-joyed by others such as clean air, beaches and forests. The existence value is the utility that consumers derive from just knowing the public good exists. A way of measuring that utility would be to measure the willingness to pay or the contingent value assigned if the public good were to disappear. Option values reflect what current generations wish to bequeath for future generations to inherit. They imply both an ethi-cal commitment to sustainabilty for the children of our children and in a shorter time-frame the maintenance of options to solve current prob-lems. Examples of option values have been devel-oped for forests relating to biodiversity and the search for cures of cancer and AIDS-"if forest were to disappear then the options to find such cures will vanish" or "as long as the forest is pro-tected there is the option to find the cure".

Direct Effects Valued on Conventional Markets Some methods are directly based on market prices or productivity. This is possible where a change in envi-ronmental quality affects actual production or pro-ductive capability.

Change-in-Productivity: Development projects can affect production and productivity positively or neg-atively. The incremental output can be valued by us-ing standard economic prices where available. Loss-of-Earnings. Environmental impacts can signifi-cantly affect human health. In theory, the value of health impacts should be determined by the willing-ness to pay of individuals to maintain their health. In practice, one uses earnings lost upon early death, dis-ease or job absence. This approach is used in highway and industrial safety, and in air pollution studies.

The "implicit value of human life" approach is reject-ed by many as dehumanizing since human life can be said to have infinite value. However, society, govern-ment regulations, insurance companies and judicial courts implicitly and explicitly place finite values on human life and health. This is a necessity reflecting limited resources to be allocated for health expendi-tures. The relatively high level of health expenditures in Cuba would indicate a high implicit value of hu-man life and health. However, one can also discern a political motivation and its attendant benefits behind it.

Preventive Expenditures: Individuals and governments invest in prevention measures to avoid or reduce un-wanted environmental effects. Environmental damages, are often difficult to assess, but historical infor-mation on preventive measures and their costs may be interpreted as a minimum value for the expected benefits that the preventive measures seek. If it is found, for example, that there is a historical pattern of under investment in Cuba for natural disaster planning and prevention; then, it can be inferred that the benefits expected from such preventive measures have been very low. This conclusion would in turn imply that the loss of human life would be assigned a relatively low value. While this possible conclusion may appear to conflict with the high political priority for health expenditures, it can be observed that the level of damages or loss of lives caused by natural di-sasters cannot be easily attributed to government in-action and thus the political cost may be easily ex-plained away.

Potential Expenditure

Valued on Conventional Markets

Replacement Cost: Simply, the costs that would have to be incurred in order to replace a damaged asset. The estimate is not a measure of benefit of avoiding damage since the damage costs may be higher or low-er than the replacement cost.

Valuation Using Implicit (or Surrogate) Markets

Sometimes one must use market information indi-rectly. Approaches to be considered are the travel cost method, the property value approach, the wage dif-ferential approach, and uses of marketed goods as surrogates for non-marketed goods. Each technique has its particular advantages and disadvantages, as well as requirements for data and resources. One must determine which techniques might be applica-ble to a particular situation.

Travel Cost: This approach measures the travel cost (and time involved) which reflects the willingness of consumers or users can serve to measure the benefits produced by recreation sites (parks, lakes, forests, wilderness). It can also be used to value "travel time" in projects dealing with fuelwood and water collec-tion.

Property Value: This valuation method is based on the general land value approach and can determine the implicit prices of certain land areas. The property value approach can help analyze willingness to pay for properties with different pollution levels and infer the implicit cost of pollution. The method compares prices of houses in affected areas with equal size and similar neighborhood characteristics elsewhere in the same metropolitan area.

Valuation Using Constructed Markets

Contingent Valuation: When society's preferences as revealed in market prices are not available, the con-tingent valuation method tries to obtain information on individual preferences by posing direct questions about willingness to pay. It basically asks people what they are willing to pay for a benefit, and/or what they are willing to accept by way of alternate compensa-tion to tolerate an environmental cost. This process may be achieved through a direct questionnaire/sur-vey. Willingness to pay is difficult to measure and de-pends on the income level of the sample subjects, and involves problems of designing, implementing and interpreting questionnaires. While its applicability may be limited, there is now considerable experience in evaluating the quality of supply of potable water and electricity services.

Artificial Market: Such markets can be constructed for experimental purposes, to determine consumer willingness to pay for a good or service. For example, a home water purification kit might be marketed at various price levels, or access to a game reserve may be offered on the basis of different admission fees, thereby facilitating the estimation of values placed by individuals on water purity or on recreation facilities.

THE DISCOUNT RATE

Discounting is the process by which costs and bene-fits occurring in different time periods may be com-pared. The discount rate to be used has been a gener-al problem in cost-benefit analysis, but it is particularly important with regard to environmental issues, since some of the associated costs and benefits are very long-term or irreversible in nature. In stan-dard analysis, past costs and benefits are treated as "sunk" and are ignored in decisions about the present and future. Future costs and benefits are discounted to their equivalent present value and then compared. In theory, in a perfect market, the interest rate re-flects both the subjective rate of time preference (of private individuals) and the rate of productivity of capital.

Higher discount rates may discriminate against fu-ture generations. This is because projects with social costs occurring in the long term and net social bene-fits occurring in the near term, will be favored by higher discount rates. It is often argued that discount rates should be low-ered to reflect long-term environmental concerns and issues of intergenerational equity. However, this would have the drawback that not only would eco-logically sound activities pass the cost-benefit test more frequently, but also a larger number of projects would generally pass the test and the resulting in-crease in investment would lead to additional envi-ronmental stress.

Many environmentalists believe that a zero discount rate should be employed to protect future genera-tions. However, employing a zero discount rate is in-equitable, since it would imply a policy of total cur-rent sacrifice, which runs counter to the proposed aim of eliminating discrimination between time periods-especially when the present contains wide-spread poverty.

In the case of projects leading to irreversible damage (e.g., destruction of natural habitats, etc.), the bene-fits of preservation may be incorporated into stan-dard cost-benefit methodology using the Krutilla-Fisher approach. Benefits of preservation will grow over time as the supply of scarce environmental re-sources decreases, demand (fueled by population growth) increases, and possibly, existence value in-creases. The Krutilla-Fisher approach incorporates these increasing benefits of preservation by including preservation benefits foregone within project costs. The benefits are shown to increase through time by the use of a rate of annual growth. While this ap-proach has the same effect on the overall CBA as low-ering discount rates, it avoids the problem of distort-ed resource allocation caused by arbitrarily manipulating discount rates.

CONCLUSIONS

In order to achieve economically sustainable manage-ment of natural resources and environmental protec-tion, one must effectively incorporate environmental concerns into decision making through the EIA pro-cess.

This presentation has reviewed concepts and tech-niques for economic valuation of environmental im-pacts within EIA procedures. The process of internal-izing these environmental externalities can be facilitated by making rough qualitative assessments early on in the project evaluation cycle-the advan-tages of which would include:

- early exclusion of alternatives that are not sound from an environmental point of view;
- more effective in-depth consideration of those al-ternatives that are preferable from the environ-mental viewpoint; and
- better opportunities for redesigning projects and policies to achieve sustainable development goals.

In order to fully reflect society's values and preferenc-es about environmental values, non-market methods of estimation can be used and will enhance commu-nity participation through well designed and admin-istered questionnaires and surveys. Research and training about EIA and embedded economic analysis methods is needed in Cuba. As developing countries learn to successfully apply EIA methodologies the goals of sustainable development will become more attainable.

4

TRANSBOUNDARY ENVIRONMENTAL IMPACTS AND PUBLIC PARTICIPATION

INTRODUCTION

The Convention on Environmental Impact Assessment in a Transboundary Context (the ESP00 Convention), was signed by the European Community in 1991. This initiated changes incorporated into the EIA Directive 97/11/EC and the IPPC Directive 96/61/EC, creating provision for cross border public consultation. Article 7 of the EIA Directive requires member states to send information 'as soon as possible' and not later than when its own public is informed. The affected state must be sent a designation of the project together with any available information on its possible transboundary impact, as well as information concerning the nature of the decision to be taken. If the affected state wishes to participate in the assessment process, it must also be sent the application submitted by the developer under Article 5, together with the information relevant to the assessment procedure. All the above information must be made available to the public in the affected State, should the latter decide to participate in the assessment process.

Within a reasonable time scale, submissions must then be accepted by the competent authority in the Member State, in which the development is proposed. The notifying Member State under Article 8 must take the transboundary consultations into account in making a decision, and under Article 9 must inform the Member States consulted under Article 7 of the decision made. With regard to the decision there is no compulsion to notify the public in the affected member states, even if they make submissions to the transboundary EIA process. (Macrory, R, and Turner, S., 2001). Article 17 of the IPPC Directive, drafted in parallel with the 97/11 Directive is less explicit with regard to the provision concerning cross border public consultations and it is proposed to amend both to bring them in line with the Aarhus Convention.

With regard to transboundary EIA, these proposed amendments will ensure equal rights of consultation for the public whether they live in the notifying or affected Member States. However, in the case of the IPPC permitting process, the public of the affected state will remain entitled only to access the IPPC application itself (Macrory, R. and Turner, S. p.14). The Water Framework Directive is still the only measure that includes not just transboundary consultations but also power sharing (Article 14(i).) Regardless of the above, it would appear that Ireland had failed to implement fully the 1985 and 1997 EIA Directives with regard to Transboundary consultations, as ruled by the European Court (Case c-392/96 (1999) CMLR 727). The Planning and Development Act 2000 whilst it empowers the Minister for the Environment to make regulations regarding transboundary EIA, it 26 does not give rights of consultations for the Irish public nor that of other Member States (Macrory, R. and Turner, S. p.17). As can be seen from the above, there is a great lack of clarity on the issue of consultation with regard to transboundary environmental issues.

PUBLIC PARTICIPATION PRACTICE

Programmes to encourage the public in EDM processes often have multiple purposes, and no simple formula exists for success. In each case a programme has to be designed to fit the particular combination of process, proponent and public. A number of case studies are given by Acland, A. (2002. pps 25-30), Ortolano (2997, pps.409-417) and the European Commission (COM (2002) 704 final. pps 62-72) and are illustrative of this point. The proponent of an EDM process be it a public agency or private enterprise must be clear as to their aim in involving the public.There may also be different aims at different stages of a process, for example within the EIA process from the elicitation of values relevant to site selection at the project design stage, to the optimisation of trust and credibility at the monitoring stage (Petts, J. and Leach, B. 2000. p.20). In order to put together the aims with the methods it is necessary to have a plan:

Planning for Public Participation

In order to make a coherent plan it is necessary for the practitioner to establish the objectives of the process, based on the overall aims. It is then important to identify which public(s) the participatory process is trying to include.

Objective Setting

Clear objectives in planning a public participation process are essential, and ideally they are delineated by consultation with a representative group of stakeholders. A community assessment process, involving community interviews (Marxen, J. 2001. 6-4), would provide a good basis for a plan. The California Department of toxic Substance Control, in its 'remediation and permit procedures' involves a public participation specialist to review the findings of such interviews (Marxen, J. 2001. 6-20).

It is equally essential that all those involved, both active and nonactive participants understand the objectives, as well as the limitations on what can be achieved. The connection between process and outcome must be clear to all. Ongoing process evaluation is essential to keep a clear focus on the objectives (Cuff, J. 2001. p.75). The two basic reasons for identifying and classifying objectives are; that objectives change over the various stages of a process, and that some participation techniques are better than others for achieving particular objectives (Canter, L.W. 1996. p.593).

Important Factors in Objective Setting

Apart from the above, there are a number of specific factors that underlie the setting of objectives, that will need to be considered at the outset, and will be present throughout the EDM process.

Environmental Protection The main purpose of EDM, and thus the main purpose of public participation in EDM is to achieve protection, conservation, and wise management of the environment. This can only be achieved if the proponent properly collects (and acts upon) evidence, opinions and perspectives from all the interested or affected stakeholders, who are to be fully involved in the EDM process, and from the earliest opportunity. Legal Requirements Appendix 1 shows the major current legislation relating to public participation in EDM in Ireland.

Opportunities to participate are present throughout the process of making a Development Plan, a Waste Management Plan and a River Basin Management Plan, however, most of the other processes are of the 'report and comment' type where the public's reaction is sought to an already drafted document. As discussed elsewhere, this situation is changing with the Directives that are emanating from Europe implementing the Aarhus convention. The legal processes, their limitations and the concomitant responsibilities should be made clear to all the participants at the outset. Proponent Objectives vs Public Objectives. The various stakeholders in an EDM process can have very different expectations of that process. The proponent may be only conducting a public relations exercise or fulfilling a legal requirement, having a final decision already made. The participants may want an influential voice in the EDM process and expect their input to be valued and included.

The proponent, seeking a mandate for a particular project, may seek to defuse opposition through consensus building. Conversely, the individuals or groups involved may be largely uncompromising on their expected outcomes. The proponent may be restricted by time and/or money, whereas the public participants may be highly motivated by value judgments made on a particular issue, and prepared to commit much personal time and energy to it. Participation can then be viewed as a tool that the public and the project proponent or planmaking agency can use for their own purposes. Two Way Communication. The development of a two-way process of communication between the regulatory authority and the public is essential in order to identify public concerns and values; inform citizens of authority's plans; and inform the authority about alternatives and impacts.

Experts vs. Public. Accepting that the role of regulatory authorities is to serve the public interest, the challenge is to integrate technocratic and democratic contributions, when addressing complex issues (Lein, K.L. 2003. p.57). Dearing and Rogers (1996) state bluntly that, "scientific research results do not play a role in the agenda setting process". The social construction of reality is what gives issues their importance, such that the building of a waste incinerator creates much stronger reactions and media attention that regular back yard burning of waste, regardless of the relative levels of dioxin release. Levels of Participation. The setting of objectives is constrained by the level of participation being allowed in any one process. It is useful to look at "Arnstein's ladder" (Ortolano, L. 1997. p.417) /and for the agency or proponent to analyse its motivations in establishing the process.

1. Manipulation
2. Ther Non-Participation. These levels assume a passive apy audience, which is given information that may be partial or constructed.
3. Informing
4. Consultation Degrees of Tokenism People are told what is about to happen, what is happening now, or what has already happened.
5. Placation People have a voice, but no power.
6. Partnership People's views have a small influence on decisions made by traditional power holders.
7. Delegated Power Degrees of Citizen Power People negotiate with traditional power holders, agreeing rules, roles, responsibilities and levels of control.
8. Citizen Control Some power is delegated.

Arnstein represents the levels of citizen participation as the 8 rungs of a ladder and groups the rungs into three groups. The continuum stretches from going through the empty ritual of nonparticipation, to having the real power needed to affect the outcome of the process. Arnstein describes this first category non-participation as tactics whose real objective is "to enable power holders to educate, or cure the participants". In the degrees of tokenism Arnstein argues that when these are "proffered by power holders as the total extent of participation, citizens may indeed hear and be heard. But they lack the power to ensure that they are heeded".

The final category, citizen power, involves citizens-agency partnerships such as those in the Bantry Bay Coastal Zone Charter and finally programmes in which citizens are in control, or can veto agency decisions.

The further the process is from centralised power, the further up that ladder it is possible to go. Thus, at the European Union level, public participation has reached its peak at the level of consultation, whereas there are now many local partnerships under Agenda 21. Generally, however, regulatory authorities are not empowered to go beyond the consultative level in public participation. Wilcox, D. (1994) adapted Arnstein's Ladder to give four categories of participation methods, and these present an easier model for the purposes of this study, as will be seen later. They are:

1. Education and information provision
2. Information feedback
3. Involvement and consultation
4. Extended involvement

These categories are explained in Section 5.4. To enable the selecting of objectives, Canter, L.W. (1996) looks at a number of systems for categorising them. One such gives three general types of objectives, which further divide to give seven more specific, viz: Public Relations-Legitimizing the role of the proponent or regulatory body. Developing confidence and trust in the stakeholders. Information-Diagnosis of problems and needs. Development of alternative solutions. Evaluation of consequences and alternatives. Conflict resolution-Consensus seeking. Depolarizing interests. One international body, The Institute for Participatory Management and

Planning (Connor, D.M. 2003) (I.P.M.P. 2003) goes much further in breaking down the objectives, with fifteen basic objectives under three categories.

The IPMP list is based on enabling a public participation practitioner, working for a regulatory authority or proponent for a project, to analyse their objectives, and then establish the best methods to achieve them. Each objective is coupled to a check-list. Viz:

- Responsibility. Establish the legitimacy of your Agency and you project.
- Maintain the legitimacy of your Agency and your project.
- Establish the legitimacy of your process.
- Maintain the legitimacy of your process.
- Establish and maintain the legitimacy of assumptions and earlier decisions.
- Responsiveness. Get to know all your potentially affected interests.
- Get to see the project through their eyes.
- Identify all potentially relevant problems.
- Generate solutions.
- Identify and classify the key issues.
- Effectiveness. Nurture and protect your credibility.
- Have your communication received and understood.
- Receive and understand information that's communicated to you.
- Search for common ground between polarized, potentially affected interests.
- Mediate between polarized interests.
- It is clear from this that one of the main objectives of practitioners is to maintain their credibility and that of the process, without which all else is in vain.

The code of ethics for practitioners, from the International Association for Public Participation Practitioners (IAP2), is also a useful general guide in setting the objectives of an effective process (IAPZ. (a). 21/5/03).

- o Purpose. We support public participation as a process to make better decisions that incorporate the interests and concerns of all affected stake holders and meet the needs of the decision making body.
- o Role of the Practitioner. We will enhance the public's participation in the decision-making process and assist decision makers in being responsive to the public's concerns and suggestions.
- o Trust. We will undertake and encourage actions that built trust and credibility for the process among all participants.
- o Defining the Public's Role. We will carefully consider and accurately portray the public's role in the decision making process.
- o Openness. We will encourage the disclosure of information relevant to the public's understanding and evaluation of a decision.
- o Respect for Communities. We will avoid strategies that risk polarising community interests, or that appear to 'divide and conquer'.
- o Advocacy. We will advocate for the process, and not for any interest, party or outcome.
- o Commitments. We will ensure that all commitments made to the public, including those by the decision maker, are made in good faith.

Identifying the Public

There are as many publics as there are different people who care, positively or negatively, about a project. Who they are may depend on their ethical, moral, interest, welfare etc. viewpoints. Other reasons the public may be affected by regulatory decisions and plans include:

- o Proximity-Pollution, Property values, Employment.

- o Economics-Landowners, House-owners.
- o Use-Amenity value, rights of way, vista.
- o Social and Environmental Issues-Justice and Risk.
- o Values-Animal Rights, Ecology, Religion.

These members of the public could be described as 'stakeholders' alongside the proponent, regulatory bodies, industry and NGO's. Wates, N. (2000 p.15) defines stakeholders as "persons or organisations with an interest, because they will be effected and may have some influence."

Whatever kind of process is chosen, it is essential to define which members of the public are stakeholders (Yosie, T.F. and Herbst, T.D. 1998. p.19). Different stakeholders are affected to varying degrees, and the greater their interest in a project, or the greater the potential impact on them, then the more involvement they should have. There are also those who could affect or impact on the project.

An individual's interest may change over the course of the participation process, and so they may opt in or out at various stages, others may want to be involved throughout the process. Some may purposefully reject participation. It is essential, however, to ensure that no group of persons is excluded, as this, whether intentional or not, will be resented, and may lead to action being taken in the legal or political spheres. The number of potential public stakeholders on some issues is practically limitless, however, not everyone can sit around the same table. The challenge then is to balance the need to consider the many view points of all the stakeholders, with the practical considerations of convening a group of individuals who have a role in making, or directly influencing, decisions.

'Tiered' stakeholder involvement is one approach used to strike that balance, and in it three categories are defined:

- o Those who want to be directly involved in the process
- o Those affected who just wish to be kept informed
- o Anyone with an interest in the project

Many stakeholders are not able to effectively participate in deliberations even when they are the most potentially affected. This may be due to poor language skills, poor self-esteem, lower educational achievement, physical handicap or intellectual disability and so on. Many remain unaware of their opportunity to participate.

The environmental Justice Programme of the US EPA, attempts to address these issues requiring "the decision makers to seek out and facilitate the involvement of those potentially affected". (US. E.P.A., 17/11/02). Participation specialists distinguish three ways of identifying segments of the public (Ortolano, L. 1997. p.405).

- o Self identification
- o Staff identification
- o Third party identification

In Self Identification, individuals and groups come forward as a result of publicity by the agency or body conducting the participation programme. These are "active" responses and should be included in a mailing list and kept continually informed of the process. Staff identification-regulatory agency personnel learn from experience the names of individuals and groups that might be interested in involvement. GIS (Global Information Systems) are being used increasingly by agencies to enable both the public and themselves to ascertain the locality of environmental problems, as well as that of the stakeholders geographically relevant to them (US, E.P.A. 2000). Other staff identification techniques include: Electoral Registers; Mailing Lists; analysis of local newspaper stories; local organisation lists; and maps. Third party identification can be achieved by asking those already involved about other individuals or groups that should be sought out. Local officials in other agencies can be asked to suggest people who may want to be involved. As each person becomes involved, they are asked to suggest others, until the response becomes minimal.

Various public involvement groups check lists exist, and can be useful as another way to ensure that no group is overlooked. (Canter, L.W. 1996. p.598) (DOELG, 1995. p.16) US E.P.A. 1999. 2-1). But it is particularly important to bear in mind that in some areas of cities, large ethnic or immigrant groups do not belong to listed groups, and are not used to interacting with government agencies. They may even be fearful of so doing. A social profile, drawn up by an applied sociologist would give a clear insight into the community from which the stakeholders are being sought. A community assessment process such as that used by the California Department of Toxic Substances Control (Marxen, J. 2001. 6-4 to 6-15) would also be an effective tool here.

Public Participation Techniques

In the Republic of Ireland, legislation has generally required the proponent of a project, or applicant for a license to notify the public of her/his intentions in this regard. The regulatory authority then takes responsibility for conducting the consultative process. However, although at present the proponents are only 'encouraged' (for example in the EPA guidelines for EIA), to involve the public at the earliest stage of project design, the Aarhus convention (Sections 4-6) 'requires' it. So, for the future, a whole range of public and private bodies will be involved in public participation, and in an increasing variety of circumstances. The following list of good practice principles (European Commission, 2000. p.239) should be considered before selection of specific methods.

General Guiding Principles

- o Plan early for public participation to establish the resources needed to support the process, and identify the limitations of finances.
- o Identify stakeholders and their legitimacy and/or representativenesss.
- o Identify the appropriate techniques to be used for each stage of the process, and provide information in a form that is comprehensible to the recipients.
- o Hold events at a time and venue to suit all participants.

- o Allow sufficient time for stakeholders to assimilate and respond to information.
- o Ensure that the inputs of stakeholders are integrated into any decisions made.
- o Ensure feedback on all issues raised.

Which methodology to use depends on many factors, but the following questions adapted from (Cuff, J. 2001. p.18) (Acland, A. 2002. 3-2) will help in deciding.

- o How complex is the issue to be discussed?
- o What processes are already in place for resolving the issue and how well are they working?
- o What processes have been tried previously and how well did they work?
- o What is the geographical scale of the issue?
- o At what levels of societal structure is the process to operate?
- o What resources are available in time, money and in-house skills?
- o Will external professional facilitators or consultants need to be employed?

With regard to this last question, apart from the costs, the employment of consultants can raise other issues. Consultants are only as good as their brief, and whilst the process may work well, (Cuff, J. 2001. p.131) the end product may be less of a success. The use of professional facilitators combined with stakeholder training in facilitation skills, early in the process, would enable a wider range of techniques to be used. It would also make the use of group-work, one of the most useful tools in concensus-seeking, more viable. (Cuff, J. 2001. p.49).

Good Practice Principles Specifically For Project Proponents (Acland, A. 2002, 3.2) include:

- As soon as possible focus on specifics, away from slogans.
- Acknowledge differences-then keep an open mind, thinking creatively to build on common ground.

- Uncertainty leads to distrust to fear to hostility, so clarify any uncertainty as early as possible. Be as open as possible and explain why, when you cannot be.
- Where legally possible, acknowledge past mistakes.
- Resolve internal issues prior to public participation.
- Address all dimensions of a situation; differences of needs and interests, behavior, and beliefs and values.
- Only make commitments you will be able to keep.
- Being open, consistent and honest in stakeholder relations is essential to building of trust.
- Flexibility in the choice of methods and use of same enables the creativity of the process.

Most important of all, it should be remembered that, stakeholder dialogue is often messy, disjointed and even chaotic at times. This is just how it should be-reflecting the real world in which we all have to operate (Acland, A. 2002, 3-2). So events should be made enjoyable. Having regular breaks for stretching, food and refreshments.

Process Planning

Many of the questions posed above can be answered during a process planning session (Wates, N. 2000. p.106) which allows stakeholders to determine the most suitable public participation methods for their particular situation.In order to do this, it is necessary for the participation coordinators to make an early attempt at identifying key stakeholders.

Record Keeping and Reporting

In order to maintain continuity and for confidence building, records of all sessions/meetings are kept and reported to all stakeholders, together with outcomes from the process. This is essential.

Public Participation Tools

What follows then is a survey of some of the tools most widely used in public participation. They are arranged in groups based on their general purpose and according to the four levels of participation outlined by Leach, S., and Wingfield, M. (1997, p.27).

Level 1 Education and information provision.

Level 2 Information and feedback.

Level 3 Involvement and consultation.

Level 4 Extended involvement.

These four levels categorise methods, according to the amount of power transfer from the responsible authority to the public. In level one, the public is passive, receiving information. In level two, feedback is requested and considered as part of the EDM process. Level three involves information, sharing and consultation, on all relevant aspects, before the final decision is made by the authority. At the fourth level, the authority can delegate power to a group of individuals to make decisions on its behalf.

Clearly, due to legislative restrictions, level four processes will be rare, except in a very local context. By contrast levels two and three will be widely used in the future. These levels are not, however, prescriptive, and some of the tools, such as interviews, can be combined with others at different levels. For a more detailed breakdown of the effectiveness of the different methods, see Canter, L.W. (1996).

The public participation tools described here are an amalgam of the author's experiences, and information from the following sources: (Canter, L.W. (1996), Gray, A.J. et al (2002) Yosie, T.F. and Herbst, T.D. (1998), IAP2(6) (2003), Ortolano, L. (1997), Connor, D.M. (2003), I.P.M.P. (2003) and Wates, N. (2000).

Education and Information Provision

Printed Materials-Fact sheets, Newsletters, Brochures, Issue papers. The guiding rule here is KISS-Keep It Short and Simple, and embellish with good visual impact. The contents should recognise the pros and cons of an issue and explain the public's potential role in, and affect on, the project. An easy response method such as a postage paid card or free-phone hotline number will enable identification of stakeholders interested in involvement.

These materials can be targeted or general, and may reach a large audience. They may not, however, be read, and are limited in what information they can convey. Connor, D.M. (2003) estimates only 10% of the necessary information can be conveyed in this format. It is also difficulty to know if you have reached your audience, unless responses are generated.

Advertisements

Whilst advertisements in newspapers or magazines can potentially reach wide audiences, they are generally expensive and limited in the information they provide. It is essential to choose the pages, days and publications very carefully. Different publics read different publications! It is, as with other printed material, difficult to know if it has been read. Some feedback mechanism should be included.

Press Releases

The potential for reaching the public is great if the issue is of sufficient interest to the media, and can include requests for feedback. There may be problems of misrepresentation and lack of editorial control. The use of publicity stunts may be of use in drawing media attention to the less exciting areas of EDM. This is a cheaper option than advertising, though local newspapers will often do features on regular advertisers more readily.

Public Displays

Exhibits in public places have the advantage of enabling a wider public to view models and displays at their own choice of time and at their leisure.

They are cheap in public buildings, but 42 to be more effective require the back up of personnel to explain details and/or provide literature. They may also need to be duplicated or travel around the stakeholder constituency. Newspaper Inserts.

To be successful an insert needs to be: of the best graphic design, in order to catch the reader's attention; in the publication on days when others are not; and preferably in the relevant local paper, where it is likely to have greater impact.

Newsletters

Although everyone will not read a newsletter it is an ongoing source of information with enough print space for some detail. It can also be used as an insert in the local paper. A newsletter also has great potential for feedback, via reply paid response forms, with for example prize draws attached.

Bill Stuffers

Any of the printed formats above could be included with utility bills, rates bills etc. Feedback possibilities can be included, these to be returned along with the bill payment.

Information Repositories

Usually, the regulatory authority makes EDM documents available at its offices and during office hours. This excludes, or makes difficult, access for many stakeholders.

Libraries, as an alternative, are locations which have hours designed for public accessibility and whose 'raison d'etre' is in part the making available of information (EC1, 2000). Information repositories are, however, not widely made use of by the public. (IAP2 (6), 2003).

Site Visits

Small groups of stakeholders can learn more from first hand experiences than from any other means. It may, however, be difficult to identify a suitable site, that replicates all the issues faced by those in the EDM process.

Video

A production controlled by the proponent or the regulator may be seen as propaganda, but if produced with stakeholder involvement may be a good substitute for site visits, and can be shown widely over the stakeholder constituency. Whilst it is expensive to produce, a professional and credible video, once made, is not too expensive to reproduce, and so can be distributed to community groups, NGO's, employees etc.

Independent Technical Documents and Expertise.

Such documents could give credibility to a process, though they may be incomprehensible to the general public unless a non-technical summary is given. Public access to experts builds credibility and may help resolve disputes over disputed 'facts'. The reality exists, however that these technical experts may contradict information provided by the project proponent.

Information Feedback

Public Meetings

This widely used method is described by (Connor, D.M. 2003) as, "the last of the blood-sports, and like the others, should be banned". Public meetings usually involve a speaker or panel giving a presentation, followed by a question and answer session. Sometimes these are broken down into smaller group discussions returning to plenary sessions.

Gray, A.J. et al, (2002) suggest that public meetings offer an opportunity for stakeholders to meet, as well as demonstrating the willingness of the proponent to meet the other interested parties. However, they go on to say that they are one of the most complex and unpredictable methods.

They can also be intimidating and can be hijacked by vocal individuals or interest groups. Badly handled, they can entrench opposition to a project, and division within a community. As a means of sharing information with those able and interested to attend, it is reasonably cost effective if run by well-trained staff.

Public Hearings

These formal occasions have a hearing officer, and a stenographer. Formal presentations from the public are received. This rather rigid format should according to (IAPZ (b) 2003) be avoided if possible, as it does not foster constructive dialogue and can perpetuate an 'us versus them' feeling, with little dialogue between participants. This format enables the public to speak without rebuttal, which is positive or negative depending on perspective.

The Internet

The use of the internet as a means of making available large amount of regularly up-dated reference material has been seen by many as the way forward in information provisions. The development of the European Commissions 'single access point', 'Your Voice in Europe' (Europa, 2003) is a good example. This portal site provides access to: every information source in the EU; consultations, where opinions are solicited on a wide range of policy-making and rule-making decisions; discussions with the EU leaders on-line; scrutiny of the working of the EU, with the possibility of giving anonymous feed back; and useful links to other sites and ways the public can be heard. There are, however, pitfalls in relying too heavily on this low cost innovative tool, and its potentially global audience. (Hacklay, M. 2002) found that only 10% of the total population of the U.K. had daily access to the Internet, with only 3% of those coming from the lowest income group, and 48% from the uppermost group. Hacklay also found, that only a small minority of the population, having access to the Internet, regularly entered the environmental information sites. This minority came mainly from the educated middle classes. He also concluded that the traditional media were the most used sources of information on the environment.

The Central Statistics Office, 'Fourth Quarter 2000 Survey' (C.S.O., 2001) showed 417,000 households in Ireland owned computers, with 262,700 having internet connections. This latter figure representing 20.4% of all households was the quadruple of the 1998 figure, and is likely to be much lower than that existing in 2003. Based on these figures, and Hacklay's findings, it is clear that caution should be used in relying on the internet as a means of communication.

Free-phone Telephone Lines

Well advertised, these phone lines can be used for the public to obtain information, ask questions or make comments about proposals or issues. The person on the line must be dedicated, well trained and have access to the information themselves, with a guaranteed call back time for complex answers that require research.

This shows transparency and allows for up to date information to be easily available. Good examples of this type of service are: the Europe Direct, Free Phone 00800 678910; and the ENFO information line, Lo-call, 1890 200191.

Interviews

One to one meetings with stakeholders provide for a two-way flow of in-depth information and can be used to evaluate potential members of community advisory/liaison groups. Interviews are very useful for developing and refining public involvement and consensus building programmes. However, this can be a very time consuming process, if multiple interviews are scheduled. Closed meetings can also be treated with suspicion, especially by those stakeholders, which do not trust the independence of the process.

Surveys

Face to face surveys, conducted in a 'random' manner, can gather information from people who would not, or could not, become involved in other ways. As with all data gathering it is essential to have a clear purpose in asking the questions, and a use for the results achieved. This is an expensive method, but confidential surveys can elicit more candid results.

Response Sheets

As mentioned earlier, leaflets etc. can have reply paid cards attached, and these can be used to gather information on public concerns and preferences. These would generally have options for being included on a mailing list. There is, however, no statistical basis to the results, as the method is only as good as the distribution, presentation and in some cases the original mailing list. Results can be easily skewed.

Random Postal/Telephone Surveys

Research of this nature needs professional planning in order to attain statistically useful results, and can be labour intensive and expensive. The response rate to telephone surveys is much higher than to mail shots, but they are also more expensive to operate. That said, these surveys would generally reach a much wider cross section of the public, including those unlikely to attend meetings. The statistically tested results produced would also be more persuasive with all the stakeholders. These surveys are generally used for attitudinal research.

Internet Surveys

Precision in setting up a site for information exchanges is essential, otherwise chat rooms can generate far more input than can be looked at. The response rate is generally high but the user profile is limited by the 'digital divide,' as mentioned earlier, and so access is limited. There is of course no geographic control over the reach of the poll, it can be easily skewed and is very labour intensive.

Deliberative Polling

To measure informal opinion, randomly selected members of the public are brought together to listen to expert opinions and their own views on an issue. They are then confidentially questioned about their opinions. The result provides a representation of the publics considered judgment. This can take 2-3 days, involves up to several hundred people, requires the use of trained facilitators, and as a result can be very costly. Television cameras can be used to document the proceedings if the issue is important enough, and a resulting documentary used to air the debate more widely.

Teleconferencing

This enables face to face questions and answer sessions. This is useful where the stakeholders find it difficult to travel. This can be an expensive process, but also an inclusive one. Access to teleconferencing systems, may however be limited.

Presentations to Community Groups

At their regular meetings, social and civic clubs often invite guest speakers. The key here is to keep it short and simple (KISS), use visual aides and any other props that enable a lively presentation. The audience here may not have attended other functions, and this enables an expansion of mailing lists. The presenter has control of the information, but feedback can also be elicited. As the guest, the presenter has no input into who is invited to attend, and it is possible that project stakeholders may not be present..The audience may, however, become stakeholders.

Care must be taken to ensure the presentation is not too technical for the audience. Once prepared, the presentation can be used many times in the same format, adapting as necessary. Expert Panels. Members of local and national media are invited to publicly interview an 'expert' panel, followed by public participation. Efforts should be made to involve the stakeholders in choosing a neutral moderator, and setting the ground rules for the meeting. This format encourages media education and gives the opportunity to dispel scientific misinformation, alongside a balanced discussion of the key issues. It does, however, require substantial organization, and may even enhance public concerns by increasing the visibility of the issues in an uncontrolled debate. The experience of the author is that, like public meetings, these are not too productive for the end game of environmental protection.

Field Offices

Where the stakeholder constituency is over a wide geographic area, the use of shop fronts in various accessible and much frequented locations, on stalls in enclosed shopping precincts and at festivals etc. can provide for a broader public outreach. The provision of well trained staff, brochures, video and internet access points will all add to the effectiveness. Such traveling units can be brought to schools, third level institutions etc. This is a relatively expensive method, but provides opportunities for positive media coverage as well as providing for responsive, ongoing and focused communication.

Informal Small Group Meetings

These would usually be in a person's home within the neighbourhood, and involve a polite and appreciative member of staff. A coffee morning gives the relaxed setting conducive to effective dialogue. Clearly the numerical outreach is small, but the meeting being held on the communities own 'turf' and at their invitation creates a new relationship of equals.

Involvement and Consultation

Workshops

These small public meetings (9-24 people) are designed to involve the participants in an active way. Wates, N. (2000) shows a typical format. They are a powerful tool for in-depth discussion, and with a good facilitator; the views of all those present will be heard and considered. It is also a good format for decision-making. Workshops can be used to promote involvement and discussion at larger presentations, as well, with the attendees splitting up into small discussion groups, each with a facilitator, and reporting back to the main body.

This prevents domination of proceedings, by individuals or organizations. Hostile participants, by contrast, may see it as 'divide and conquer'. Workshops humanize project team members, promote public ownership of the process, and can be targeted at particular stakeholders. Before workshops commence, it is essential to know what is to be done with the inputs received. Workshops can also be used to 'thrash out' issues raised in other formats.

Focus Groups

These are a quick means of gauging what is likely to be the public reaction to a proposal. Involving a group of 15-25 invited participants, properly facilitated, this method can give a detailed understanding of people's perspectives, values and concerns. Selection of the participants so as to involve a good cross section of stakeholders is critical. Such a forum could be a one-day event, repeated as necessary. The participants' needs for information etc. would have to be serviced. Conflicting viewpoints can be resolved, and personal respect engendered, in this as in all other small groups. If well planned and facilitated, Focus Groups are relatively inexpensive to organize.

Open House

Like any open day, this is the opportunity for the proponent, or regulator, to give a broad and detailed account of the project.

The public are invited to a venue, preferably a neutral one in their own community, any time during the opening hours of 2-9.00 p.m., to enable maximum attendance. The individual is greeted at the door and the layout of information is explained. The members of the public then visit a number of 'stations', each illuminating a particular point and viewable by 6-10 people at a time. Each station is staffed by a resource-person, who can answer questions, and explain the exhibit. Visitors are asked to fill out comment sheets.

To make this a worthwhile exercise requires good publicity and the preparation of staff and materials, as well as time spent on the day. This makes it an expensive method, and one which, without added attractions, will not get much media attention. Conversely, it enables small group/one to one communication. It also allows team members to share information, and when needed, to answer visitors questions. The open house builds credibility. Open houses can be operated on the internet. Such open houses are cheap, accessible at all hours and not geographically limited. However, the 'digital divide', and the openness of the process make it difficult to use the inputs in any authoritative way, except as a source of ideas.

The Delphi Method

Following mail-shots, directed or random, with reply-paid questionnaires, the responses are collated, and then copies of the collated responses are sent to all those who responded for further comments, and so on. This enables consensus building, without the necessity for meetings and can also be done using combinations of telephone and post, or, with the caveats previously mentioned, by e-mail. It is a relatively cheap way of assessing the major issues relating to a project. The same methodology can be used to achieve expert consensus over contentious facts.

Extended Involvement

Citizen Juries

Essentially an informal inquiry, this method involves 15-20 ordinary members of the public selected to represent a cross section of the community. They are brought together to consider a particular issue and receive evidence from expert witnesses, whom they can cross-examine. At the end of the process they produce a report setting out the views of the jury, including differences of opinion. Although its findings are non-binding, and the jury has no legal standing, the commissioning body must follow recommendations or explain why. In this way, issues can be considered in depth and in a credible way by, 'ordinary citizens'. Flaws can be pinpointed and public reaction gauged, and in a fairly short period of time. It is essential to plan the jury well, being clear, as to the resources necessary, and about what will be done with the report and its recommendations. This is a resource intensive approach, requiring the payment of jurors, a facilitator, expert's fees etc.

Advisory Groups

These can be statutory as in the case of the EPA (Section 27 of the EPA Acts, 1992) or not, and may be granted powers ranging from making recommendations to exercising leverage over final decisions. They are usually a small representative group of stakeholders who meet to discuss issues within a particular remit. Their existence is long term and not issue specific, and requires a large commitment from the participants and more resources than some of the other methods. Rules and responsibilities must be clearly defined in advance, and a consistently credible, transparent process, with third party facilitators, is essential. Potential committee members must be interviewed before appointment and the necessary support structures put in place. Participants gain a good understanding of the issues, processes and other viewpoints leading towards compromise. There is, however, the risk that the public will see the committee as too 'cosy' with the sponsor. The sponsor must be willing to take the groups advice seriously, and accept when consensus cannot be reached.

Task Forces

Generally issue-specific, a task force, made up of independent experts and stakeholders can identify, evaluate and solve particular problems. Such a group has great credibility and provides constructive opportunities for compromise. Time and labour expensive, task forces may not come to clear decisions or reach concensus.

Planning for Real

This innovative method was designed originally for communities to use in urban land use planning, but has been adopted for other situations. A good example of this latter was 'Planning for Flood Plains' (Cuff, J. 2001. p. 2.1.2.) a river basin management project. The process contained in the registered brand name 'Planning for Real' focuses on the construction of models and the use or priority cards so reducing the need for good verbal or written sills. This empowering process enables the community to identify their own aspirations and issues, bearing in mind the constraints of the real world (Wates, N. 2000. p.100). Dedicated staff time is needed and the complex 3.D models are expensive. Community Visioning. Most suited to developing general plans for a site area or organization, this has been much used by UK Local Authorities for Agenda 21 processes. This is a pre-planning process designed to collectively create imaginative shared visions for what the future could be in a community. Because there is no control over the outcome, the project proponent must be clear that considerable power has been handed to the community.

That said, this process is likely to lead to strong commitment to the outcome from all participants. Visioning can also be done via the internet, but there is no control over who makes the inputs, and restricts the process to internet users. A variation on the process is the 'Future Search Conference' used to focus on the future of an organization, a network of people, (IAP2 (b) 2003) or a community. As with most of these techniques, it is essential to use a trained facilitator. This method can involve hundreds of people simultaneously and each one is an 'expert'. A 2-3 day meeting, this may prove logistically challenging. It may also prove difficult to gain complete commitment from all stakeholders.

Evaluating the Process

Evaluation has been defined as 'a process of assessment which identifies and analyses the nature and impact of processes and programmes' (Interact, 2001). The evaluation of public participation is, however, a relatively rare activity (Gray, A.J. 2002. p.46). Yet it is an essential feedback mechanism during the process for the stakeholders, and post process for the funders, proponents, and stakeholders, to see, if the process is going well, or whether it was worth the effort. As a learning process for future processes, clearly, record keeping and assessment are essential. Clearly then, it is not just the end result that is of importance in this assessment, but the journey to reach it.

Most of the international evaluations use 'fairness; and 'competency' as underlying criteria (Webler; T. 1995), and most rely on 'generic' expert criteria. But Barnes, M. (1998) suggest that the evaluation process itself has to adopt participatory techniques if the criteria for assessment are to include those considered important by all the stakeholders. Looking at the assessment of participatory processes for the UK Environment Agency, Petts. J. (2000) uses some 12 effectiveness criteria.

- o Clarity of objectives-is the connection made between purpose process and outcome?
- o Clarity of legal process-including limitations of outcome.
- o Consensus on Agenda and Procedures including choice of facilitator or chair person.
- o Input into assessment process.
- o Representativeness.
- o Inclusivity-are all barriers to participation minimized, and are non-participant stakeholders kept informed?
- o Deliberation-is dialogue open, inclusive, informal, detailed enough, educational and constructive?
- o Capability-are sufficient resources in time, money and staff training available?
- o Sound learning-is the process enjoyable and informative? Does it lead to increased consensus?
- o Actual Decision Responsiveness-is the process outcome included in the final decision made by the relevant authority?

- o Trust Enhancement-has the process increased trust between proponent and stakeholder for future participatory processes?

Perhaps a more useful tool, however, is that developed for the EU Life-Environment Project, 'Wise use of Flood Plains' (Cuff,J.,2001,pps 75-81). This could be used for an individual public participation method or for a whole public participation process. Figure 2. Cuff's Circle. Summary framework for evaluation of participatory processes. (Cuff, J., 2001) The categories are adapted here, where necessary, for a more general use by the author. The circular process, with 'outcomes' as its centre, and about which all else revolves, is a useful visual to maintain focus throughout the process. Objectives-These could be the objectives of the proponent and/or the objectives created as part of planning for the process. The two sets of objectives could be compared to see if participation objectives complement strategic ones. Did the objectives change over time and were they realistic. To what extent were objectives met. Contexts-Is the process connected to: for example rule making, long term planning, licensing, land use planning or a combination of two or more, e.g. a planning application with an EIS and an IPPC Licence process all on the one project? If so, how did the different processes interlink?

Other factors could be political (e.g. local elections) or geographical (e.g. a long powerline where jobs depend on it at one end, but people along its route resent it). Increased pressures on a process can be due to a need to act, brought about by external factors, e.g. flooding of the Somerset Levels (Cuff, J. 2001. p.77). Levels of Involvement-Using Arnstein's ladder, methodology can be assessed and explanations given for the level of power transfer used, in the light of the submissions made. Inclusivity and Stakeholder Identification-This should also be measured in the light of the objections. Who was left out, or included? Were there complaints about this? Was a special effort made to include socially excluded groups? Were events timed to suit stakeholders? Was there a reluctance to participate? if so, why? How many, and what type of people actually took part?

A format. that could be used to evaluate individual participatory methods or processes. Scoring: 0= went badly, did not meet objectives, 10= all objectives met, total success. Participatory Methods/Processes-This assessment should be for each method used, and an objective set for each at its onset. The use of an H Diagram gives a good picture of the pros and cons of an event. All aspects of the methodology should be scored from 1-10 to show strengths and weaknesses, and recommendations for the future use of same. Innovation-Any new methods could be shown and assessed here, for possible use in the future.

Commitment to using Participation Process Results-Was it clear at the outset what limitations the participation process had in influencing the decision-making? What actual use was made of the process outcomes? Were participants informed of the use of their inputs? If so, when and how? Inputs-A break down of the costs of the process into individual events, and then into their constituent costs, is needed to assess the value-for-money represented by the different methods. Was lack of finance an obstacle? Was the use of consultants value for money? Outputs-This is a measure of the number and quality of outputs and their effectiveness, e.g. posters, exhibitions, questionnaires, etc. Outcomes-These stem mainly from 5 and 9 (figure 2, above) and are measured against the baseline of the pre-process situation. Firstly, there are the strategic outcomes. Has a decision/recommendation been made, and is it based on a consensus? Secondly, there are the cultural outcomes. For stakeholders these include:

o Change in the attitude to participatory processes.

o Ownership of the process o Increased trust in the proponent

o Changes in values or attitudes

For organizations:

o Are they more transparent?

o Are they more at ease with public participation?

o Do they see the process as worthwhile?

There will be other outcomes in the many circumstances in which this assessment process could be used.Wates, N. (2000. p.100) describes how to organize and run an effective 'review session' involving all stakeholders.

Conflict Resolution

Having pursued a participatory programme what can be done if the stakeholders cannot reach concensus? The use of strategies known as ADR (Alternative Dispute Resolution) and based on 'joint gains' problem solving, mediations, facilitation and consensus building offer promise for many cases (Adler, P.S. et al., 2003). Bingham, G. (1986) showed that 78% of the 161 environmental disputes studies, which used mediation-based ADR, ended in agreement amongst the parties. It is, however, not a panacea, and each case should be judged on its merits. The US EPA uses the definition of ADR in the Administration Dispute Resolutions Act of 1996 "any procedure that is used to resolve issues in controversy, including but not limited to conciliation, facilitation, mediation, fact finding, minitrials, arbitration and the use of ombuds or any combination thereof. All the methods listed above involve a neutral third party who assists others in designing and conducting a process aimed at reaching agreement. All aspects of ADR are voluntary, including the decision to participate, the type of process used, and the content of any final agreement. The ADR process and practices are beyond the scope of this study and generally would only be undertaken using specialist EDM mediators or facilitators, however, it is worth looking at the underlying principle. The conditions, under which ADR is useful, and most likely to be successful are listed by Otolano (1997. p.417) as:

- A high uncertainity in timing and outcome. Stakeholders who perceive other quicker or more certain routes to achieve their objectives all have little motivation to become involved in ADR.
- Equivalent distributions of power. Here power is described as the ability to prohibit an action desired by one or more adversaries.

- o Mediation is more promising than BATNA (Best Alternative to a Negotiated Agreement). As long as the prospective outcome from negotiation is more attractive than BATNA, a stakeholder has an incentive to continue negotiating. If the prospects for success for negotiation drop thán the BATNA might be the courts, lobbying or direct action.
- o Disputing parties are at an impasse.
- o Litigation is ineffective.

Ortalano, L. (1997, p.411) goes on to describe the ADR techniques known as 'principled negotiations'. This is characterized by four main points, which are a useful insight into problem solving during any EDM process.

- o Separate the people from the problem
- o Focus on interests not positions
- o Invent options for mutual gains
- o Use objective criteria for evaluating the agreement.

Implicit in these is the recognition of the four main types of conflict delineated by Creighton, J.L. (1981. pp 454-455).

- o Cognitive conflict-differences over the facts.
- o Values conflicts-a dispute over goals, for example a dispute over whether a particular outcome is desirable (undesirable) or should occur or not.
- o Interest conflicts-Cost and benefits are rarely evenly distributed so that some may have a greater interest in a particular outcome. It is possible then to have agreement on facts and on values but still disagree on interest.
- o Relationship conflict-Communication involves both facts and relationships (how much a person is valued, accepted etc.). The EDM process itself may favour those with power and money and so communicate a lesser value to other stakeholders, leading to conflict on emotionally motivated grounds other than facts, values or interests.

CAPACITY BUILDING

Defined by Wates, N. (2000) as "the development of awareness, knowledge, skills and operational capability in order to achieve a purpose", capacity building is an essential part of the development of public participation practice, and it is necessary to build capacity amongst all the stakeholder groups in society.

Agenda 21, Chapter 36 set out the broad proposals for promoting education, public awareness and training which it sees as meeting capacity building needs. It promotes the inclusion of environmental and human ecological considerations in all aspects of education and decisionmaking.

'The Access Initiative' (Coyle, J. et al. 2002) in a study of nine nations found that: lack of government capacity constrains public access e.g. through lack of staff, equipment, procedures and training. Laws and procedures relating to access are often new, and the necessary training has not been done to facilitate their enactment. South Africa was seen as the only country, of the nine studied, with an extensive programme building staff capacity. Lack of capacity in civil society was also seen as a serious impediment in all the countries studied.

The Environmental Law Institute (Breggin, L. and Hallman, H. 1999) looked in detail at Building Capacity to participate in US EPA activities and pinpointed the following as essential elements in capacity building.

- o Information.
- o Technical Assistance.
- o Process Education.
- o Access to Documents.
- o Education on laws as fundamental building blocks.

The study then went on to give potential approaches to enabling stakeholders to participate, showing models of each method in action. These include;

- Independent Information Broker
- Ombudsperson

- Hotlines
- Technical Assistance Grants
- Citizen Training on Processes and Legal Requirements
- New Collobrative Participation Processes
- Increased Date Availability and Dissemination Networks
- Grants to Community Groups
- Improved Access to Documents
- Improved Mailing Lists

Capacity building in the Irish context will be looked at in the following chapters, but the following are a few examples of best practice at work elsewhere. The US EPA's 'Community Based Environmental Protection' programme is another way to create local ownership of the environment (US. E.P.A., 1999). The many Local Agenda 21 projects worldwide are achieving similar results. In the UK, through the Environment Council, training programmes for all stakeholder groups are available on stakeholder dialogue, as a tool to achieve sustainability. Information tools to enable local government employees to respond to their publics now include a one-stop internet portal. In Ireland, IBEC runs regular training programmes for people in Industry, on 'Environmental Communication', to promote better communication with the other stakeholders, regulatory bodies, the media, and the local community (IBEC, 2003).

IRELAND AND PRINCIPLE

The Access Initiative

Over the last three years the World Resources Institute has formulated a methodology to assess the implementation of Principle 10 at national level, called The Access Initiative (TAI) (The Access Initiative, 2003). This is a global coalition of civil society groups collaborating to assess and promote national implementation of Principle 10 commitments. The European Commission has now signed up to Partnership for Principle 10 (P.P.10), which commits governments to support such independent assessments, by NGO's, of national Principle 10 performance. The pilot project, reported in 'Closing the Gap' (Coyle, J. et al. 2002) involved nine nations on five continents and twenty-five civil society groups. Following this, the process was refined, and the difficult task begun of finding a methodology for assessing the implementation of Principe 10 in the many different cultural and political systems that exist. The methodology looks at:

(i) The comprehensiveness and quality of the General Legal Framework

(ii) The degree of available access to selected types of information about the environment.

(iii) The degree of public participation in decision making processes, and

(iv) Comprehensiveness and quality of capacity building efforts to encourage informal and meaningful public participation.

In order to conduct an assessment of the Irish government's performance, and so to pinpoint weaknesses, with a view to capacity building, the Access Initiative requires the building of a coalition of civil society groups within Ireland.

The formation of a coalition is necessary to get the skills and knowledge base necessary to perform the research and analysis tasks contained in a full Access Initiative Assessment. Potential coalition members might include NGO's, lawyers, academics, journalists, etc. The structure of the coalition would include a local organization, a research team with membership from across the coalition, and an advisory group of 'experts' to oversee the research and give credibility to its findings (The Access Initiative, 2003).

The work involved, with 6-12 researchers, is estimated to take 4-6 months, based on the pilot project. Clearly the degree of detailed research required to carry out a complete TAI study of Ireland is beyond the scope of one individual and of this study, but a selective look at the issues raised will give some insight into the current state of environmental governance in the Republic.

Access to Information

Law: Whilst there is not constitutional right to information on the environment, freedom of access to information is available under the Freedom of Information Act, 1997 (FOIA) and the European Directive 90/313/EC (S.I. 125 of1998) (replaced by 2003/4/EC). Access under FOIA is restricted by cost, limiting its use by people with low incomes. Emergencies:

Although a general legal mandate exists (The Seveso Directive) for dissemination of information about environmental and health impacts to the public during and after an emergency, An Taisce has repeatedly expressed its concern about the implementation of same (Lumley, I. 2003)

Monitoring: In general, monitoring of air and water quality is carried out to a high standard and information is made available about the results of same. Both air and water quality monitoring results are available on the EPA website, and are freely available to the public through ENFO. The quality of this information is of a good standard, but the data is 1-2 years old. Local Authorities also carry information on their web sites, though the range and quality varies considerably.

State of the Environment Report (SOE): The EPA, as mandated by the EPA Act 1992, produced its third SOE Report in 2000, and it is rich in data and indicators to describe conditions and trends. The report was well publicized in the media, and available in bookshops. 'Government in Focus-Key Environmental Indicators For Ireland 2002', is available on the web-site.

Facility-level Information: Readily available and timely information on the environmental impact of particular facilities is essential to enable people to protect their environment. In the USA online information regularly updated is available on the Toxic Releases Inventory (TRI) web-site. In Ireland the EPA is developing a similar system, based on GIS mapping. Local Authorities have some information on web sites, but not facility specific. Otherwise it is necessary to trawl through registers at EPA or Local Authority offices.

Participation in Environmental Decision Making

Law: The legislative and constitutional frameworks regarding participation are beyond the scope of this study, but would be addressed in an Access Initiative Assessment. Whilst public participation is allowed at an early stage in some environmental decision making processes (e.g. Development Plans) in others it happens too late to make substantive changes (e.g. the EIA process), and in others, there is no right of appeal to an independent arbiter such as An Bord Pleanala (e.g. Article 81,Planning and Development Regulations, 2001). Policy: Lead times for notification of public participation are acceptable in terms of the Access Initiative, though the author's experience would suggest that this depends on the effectiveness of the notification procedure. There is a wide availability through ENFO, the Internet, Local Government Offices, Libraries, etc. of policies, strategies, plans programme and laws.

The requirement to send development plans, planning applications etc. to the 'statutory bodies' gives a degree of external consultation by non-involved expertise, but as explained elsewhere, much environmental decision making in Ireland is of the 'report and comment' type. Project and Plans: The lead in time for public comment on draft plans is generally eight weeks, but it is five weeks for a planning application, regardless of the project size and complexity. The quality of information provided to the public and relevant authorities, particularly EIS's is very varied, and is often, in the author's experience, poor both in content and presentation. Clearly this does not allow for good decision-making. Records of local permits and other project documents are accessible at the relevant offices and increasingly on the internet.

Capacity

Law: The right of freedom of association and the right to a clean environment are guaranteed under the Irish constitution.

Government: The establishment of FOI officers in all the government departments dealing with environmental issues, and the publication of the EPA Section 15 and Section 16 reference books are some of the ways in which capacity is being built by the government with regard to access.

The presence of ENFO (The Environmental Information Office) as a one-stop shop and library for all environmental information is a very important capacity building force. ENFO also provides support for schools and materials for environmental education. ENGO's are encouraged and give some funding under European Parliament Decision No. 466/2002/EC. It is not intended to give further comment here on access to EDM in Ireland as this will be discussed later in the light of the results of the survey undertaken as part of this study.

ENVIRONMENTAL IMPACT ASSESSMENT (EIA) AND PUBLIC INVOLVEMENT

EIA is a gateway control and not in itself a permitting or authorisation process. It is a very important tool to enable good decision-making, and although its findings usually correlate to the final decision, theoretically the reverse can be true. Article 6 para 2(e) of the Aarhus Convention states that "the fact that the activity is subject to a national or trans boundary impact assessment procedure", dictates that the public concerned shall be informed, either by public notice or individually as appropriate, early in an adequate, timely and effective manner". Further, Articles 4 and 5 require each Party to provide for public participation when all options are open, whilst encouraging applicants to identify the public concerned and enter into discussions prior to applying for a permit.

Within Ireland Council Directive 85/337/EEC as amended by Directive 97/11/EC lays down the requirements for the EIA process, and is given effect in Irish Law through the Planning and Development Act 2000 (implemented by The Planning And Development Act, Regulations, 2001.), through IPPC Licensing and Waste Licensing and through the European Communities (Environmental Impact Assessment) Regulations, 1989-1999. (EIA Regulations) (Doyle, A., 2002).

In the EIA Regulations, a developer when required must submit an Environmental Impact Statement to the competent Authority. An EIS is defined in S.I No. 349 of 89 (Article 3 (1) as follows " A statement of the effects, if any, which a proposed development, if carried out, would have on the environment". Following the submission of an EIS, the public, relevant public bodies and, where legislation requires, other governments, have the right to comment.

The competent authority must take all submissions into consideration and make a decision. The decision and the reasons for taking it must then be communicated to the public and other interested parties. The whole of this procedure is the EIA. There is no publicly available written analysis of the process leading up to the decision, unlike the IPPC and Waste Licensing processes. In these, an inspector's report to the Agency, containing an analysis of the issues, is available for public comment, in the course of an objection. Essentially through, in each of these three processes it is the presentation of a report and then public comment.

The EPA by contrast in its "Guidelines on the Information to be Contained in Environmental Impact Statements"(P.7), lists public participation as one of its four basic tenets of EIA. In the same guide (para 2.3.7) it suggests, that apart from its basic statutory rights, the public could participate in the EIA process by making observations or submissions, in response to a preliminary invitation to participate in the procedural stages of scoping; evaluation of alternatives, and document review. Public involvement early in the process enables applicants to include or address concerns before it is too late.

It further asserts that, direct participation, i.e. face to face, communication is more likely to result in accurate and focused communication. In Para 2.3.6 the role of NGO's is discussed, and their input encouraged at the earliest stage possible. "Early, open and constructive engagement has frequently proven to be beneficial to both the protection of the environment and to the quality of development projects", it states. However, at present these are recommendations only, and it is the proponents decision whether or not to involve the public, apart from the review of the final EIS.

The EIA Process

The Definition: "EIA can be defined as: the process of identifying, protecting evaluating and mitigating the biophysical, social and other relevant effects of development proposals, prior to major decisions being taken and commitments made" (Saddler, B. et al, 1999) The World Bank expands this definition and describes EIA as a procedure that; "evaluates a projects potential environmental risks and impacts in its area of influence; examines project alternatives; identifies ways of improving project selection, siting, planning, design, and implementation, by preventing, minimizing, mitigating or compensating for adverse environmental impacts and enhancing positive impacts" (Klees, R. 2002).

The main aim of the process is to stimulate thinking, and encourage action, and not just the ticking off of boxes just to get another report (Verheem, R. 2002).

The Principles of EIA

The IEA (Saddler, B. et al., 1999) identifies two sets of principles, 'basic' and 'operative'. The basic set applies to all stages of EIA as well as to Strategic Environment Assessment (SEA) of policies, plans and programmes. One of these states that the EIA process should be participative; providing appropriate opportunities to inform and involve the interested and affected public, and, that their inputs and concerns should be explicitly addressed in the documentation and decisionmaking" (Saddler, B. et al., 1999)

In a review of 25 years of National Environmental Policy Act (NEPA) the Council on Environmental Quality (CEQ) (CEQ 1997) stated "the success of a NEPA process heavily depends on whether an agency has systematically reached out to those who will be most affected by a proposal, gathered information and ideas from them, and responded to the input by modifying the proposal or adding alternatives, through the entire course of the planning process".

The EIA Cycle and Public Involvement

The EIA process should be applied as early as possible in decision making for a project, providing for the involvement and input of communities and industries affected, as well as the interested public (Saddler, B. et al., 1999)

The EIA cycle shown in Fig. 4 (above) as part of the planning system and project cycle should provide for the following: (European Commission, 2000).

Screening

Screening determines whether or not a proposal should be subject to EIA, and if so, at what level of detail. Consultation with potentially effected parties here will improve understanding of the nature and significance of potential impacts using 'local' knowledge, together with that of 'experts'. The objectives here should be: to obtain a complete understanding of how the issue is viewed by all the significant stakeholders; and to identify the future levels of interest in public participation activities on this issue. Techniques could be used here from levels 1 and 2 of the participation tools list. (Eccleston, C.H. 2000. p.65) has a checklist of tasks to consider as part of the prescoping phase. Added to this should be the preparation of a public participation plan, which includes a publicly available record of the process. The use of exploratory thinking techniques such as brainstorming and snow-storming at this phase helps stimulate the creativity need to seek alternatives.

Scoping

This is a very early exercise in an EIA in which an attempt is made to identify the attributes of the components of the environment for which there is public (and professional) concerns and upon which the EIA should be focused (Singleton, R., Castle, P. and Short, D. 1999, p.103). Good scoping is the key to a successful EIA process. Public involvement here ensures that all the significant issues are identified, local knowledge about the area is incorporated, and alternatives are identified and considered. The principal objectives of public scoping are: (Eccleston, C.H. 2000, p.72)

- Identify public concerns and the expertise needed to investigate same.
- Identify alternatives to be examined
- Identify significant issues that need to be analysed, eliminating the unimportant.
- Identify problems and potential solution early in the process.
- Identify problems with the participation process and address same.
- Ensure that both the positive and negative aspects of the proposal are identified and studied.
- Identify potential mitigation measures.

A scoping information package should be put together to promote public involvement and to inform the scoping delegates. It should include:

- An invitation to participate showing how, when and where.
- A brief description of the working of the EIS process and the opportunities for participation.
- A description of the proposal, such that the objectives of the proponent are clearly laid out with maps, diagrams, figures etc.
- A description of the known potential impacts. A public notice should then be issued using any of the methods deemed necessary from Level 1 (Section 1).

As was mentioned in the section on P.P. tools, public meetings, whilst commonly used are not the best method, unless they are based on small group sessions and workshops. Public opinion surveys, citizen advisory committees or any other methods listed in Section No. 2 Simple methods may suffice for describing, synthenisizing and communicating information on the preproject environment and the potential impacts e.g. using checklists, matrices and networks. (Eccleston, C.H. 2000. p.75) (Australian EIA Network, 1996) gives a check list for planning a public scoping meeting, and warns that "outside entities that participate in scoping generally do so because they are opposed to the proposal; the remaining participants who support the proposal often do so because they stand to gain from it.

Not surprisingly numerous public scoping efforts have dissolved into sessions of frustration, dissension or outright confrontation. So it is essential to plan well and use professional facilitators or neutral moderators where debates are likely to be heated.

Alongside the general public scoping, it can be of value to conduct focus group/workshop meetings with stakeholders having special interests or expertise, in order to examine more detailed or complex issues. The results of these meetings should be made public, and so become part of the general scoping process.

Eccleston, C.H. (2000 p.249) gives a methodology for handling the potentially voluminous, or complex scoping input, followed by an internal 'responsible authority' scoping session. This can consider issues including technical ones, using in-house expertise, and in greater depth than might otherwise be possible in a public setting.

The body conducting the EIA process, together with stakeholder representatives should then produce a post scoping document, summarising the scoping process, its findings and the reasons for decisions taken in reaching the scope for the resultant EIS Implementation Plans.

Impact Assessment/Mitigation

"Public involvement here can serve to ensure that the analysis and mitigation, necessary to avoid, minimize or offset predicted adverse impacts is relevant to local concerns, and accurately reflects local value and preferences". (European Commission, 2000).

Both impact mitigation, and the evaluation of significance are complex and time consuming processes, which cannot be dealt with by large groups. It is necessary, therefore, to use stakeholder representatives to work with the experts, reviewing the finding of the experts in workshop sessions or other facilitated small groups. Regular reporting of progress in the process, to the wider publics will maintain their 'ownership' and sense of involvement.

The Evaluation of Significance

This is a difficult issue to resolve even between specialists. It is, however, essential that the public be involved and their perspective included, as the interpretation of significance occupies a fluid boundary between science and politics (Australian EIA Network 1996).

The Canadian EA system of impact assessment is based solely on scientific, credible technical and other relevant information (C.E.A.A., 2001. The resulting determination of significance must be 'objective' and reasonable so as to withstand court challenge. By comparison in the US system, under NEPA, public opinion and the controversiality of the proposal help to identify and determine significance. Ultimately, however, the relevant authority will decide, taking all inputs on board.

Evaluation of significance is subjective, contingent on values and dependent on the environmental and community context (Australian EIA Network, 1996). The intrusion of wider public concerns and social values, into the significances evaluated by scientists is inevitable, as discussed elsewhere. The challenging nature of this part of the process is, therefore, one in which the use of facilitated small groups would be essential for the more contentious issues. More information, on the technical aspects of the evaluation of impacts. is given by Lein, K.L. (2003).

Preparation of EIS

It is essential to document clearly and impartially the impacts of the proposal, as well as those of the identified alternatives, the proposed mitigation methods, the significant effects and the concerns expressed by the public and communities affected by the proposal, as well as how those concerns were addressed.

The draft EIS, following internal review by the relevant body, should be made available for public review, and should include a non-technical summary. Review of the EIS The review of the EIS should determine whether the report; meets its terms of reference, provides a satisfactory assessment of the proposal(s); and contains the information required for decision making. Involvement of the public can ensure the quality and comprehensiveness of the assessment and help to reduce any bias in the analysis.

Before the public review, the EIS should be complete in every aspect, except this final opportunity for public feedback. Any major changes resulting from this public review would probably trigger a second public review, though good public participation early on should make this unlikely.

Proposals for electronic transmission of the document along with public access through libraries, site offices, public displays, information repositories, and presentations, should give wide availability. Methods of feedback should be widely publicised. One method could be an interactive web-site page for this purpose.

All those who made major contributions to the process should receive a copy (finance permitting) or have access to a copy, along with the statutory bodies and the proponent. The draft should be available and publicised at least 15 days before any stakeholder meetings to discuss it.

The final EIS should then include the changes based on the relevant feedback in the draft, together with the responsible body's reasons for not including any other issues raised. Eccleston, C.H. (2000) gives useful guidelines on dealing with this stage of the process.

Decision Making

It is generally the role of the regulatory authority to approve or reject the proposal and, if it is approved, to establish the terms and conditions of its implementation, taking into account the EIS, and its public input.

Follow-up, Implementation and Monitoring

The participation of local representatives and NGO's in monitoring the operational impacts of a project can lead to the early identification of problems, and can foster a sense of public partnership.

Continual assessment of the EIA process

Continuous assessment of the process itself can ensure the implementation of the public participation plan, provide public transparency and strengthen the effectiveness of future EIA's.

EIA Best Practice

In its 'Directory of Impact Assessment Guidelines' (Donnelly, A., Dalal-Clayton, B., Hughes, R., 1998) the IIED (International Institute for Environment and Development) concludes that "greater attention to stakeholder involvement leads to better environmental assessment, and thus to the formulation of projects that deliver more social benefits, fewer environmental costs and greater economic and financial benefits". In looking at best practice world wide, it will not be possible to make assessments of their effectiveness within the scope of this study. However, it finds that in reality EIA guidelines and practice generally pay only lip service to real stakeholder involvement.

This is confirmed by the European and Central Asia Environmentally and Socially Sustainable Development Department (ECSSD) regional study of Eastern European and Central Asian Countries (Klees, R. 2002) where quite often, inclusive legislation is present but the reality of implementation is quite different. Constructive stakeholder involvement in EIA will rarely occur spontaneously. A proactive approach is required; starting with clear guidelines, legally enforceable by inexpensive court action; and financial assistance for stakeholders where significant costs would be incurred in order to take part in dialogue.

There are a number of good examples of proactive legislations and stakeholder support systems. Canada. Public participation in the federal assessment process is an implicit part of the Canadian Environmental Assessment Act, 1995, which was itself the result of five years nationwide consultation. The Canadian Environmental Assessment Agency (CEAA) was established to administer the Act and to ensure that opportunities for public participation are provided (C.E.A.A., 2001). The CEAA must reach out to the public for their input into screenings, comprehensive studies, mediations and panel reviews.

A public registry is established to ensure public access to records, and the responsible authority must address the need for a follow up programme to verify the accuracy of the assessment and efficiency of the mitigation measures.

A participant-funding programme, designed to provide limited funding to interested individuals and groups, is open both to those for and against a project. Its purpose is to enable effective participation at key stages of mediations and panel review.

The 'Guidelines on Environmental Assessment and Traditional Knowledge of Indigenous Peoples' (Donnelly, A. et al., 1998. p.29), formulated in Canada, draws attention to the limitations of short term scientific procedures to collect reliable and adequate information on which to base predictions. They also explore the very different perceptions of environment and development held by indigenous and non-indigenous stakeholders. New Zealand.

The Resource Management Act 1991 (RMA) states the two key principles: that decisions on environmental matters are most appropriately made by the communities most affected by those decisions, and that community participation is vital to effective resource management. This legislation is based in part on the Maori 'Kaitia Kitanga' (guardianship) of the 'Mauri' (life essence binding the physical and spiritual elements of all life). It is the first attempt world wide at an holistic legislative approach to the management of environmental resources (Ministry for the Environment, New Zealand, 2003).

The Act simplifies access for the public by putting all activities that impinge on environmental resources under the same resource consent procedures. A resource consent gives a person or organization permission to sue or develop a natural or physical resource, and/ or carry out an activity that affects the environment in some way for a stated period. Resource consent replace all the many different permissions granted under previous law. Its simple appeals structure,

Environmental Legal Assistance Fund, and its insistence that decision makers actively consider alternatives to regulation in preparing policy statements and plans (Section 32 of the RMA), makes the RMA a highly accessible process for the public to become involved with. The Arctic.

The Arctic Environmental Protection Strategy (AEPS) agreed by the eight Arctic countries lays great importance on the integration of two ways of knowing: Traditional Indigenous Knowledge and Scientific Knowledge (A.E.P.S., 2002). Experience gained in mapping traditional knowledge on the Belluga whale in ten indigenous communities on the Alaska and Russian sides of the Bering Sea, have lead to the inclusion of the use of traditional knowledge alongside general public participation as two of the important features of the "Guidelines for EIS in the Artic"(A.E.P.S., 2002)

The USA. The Council on Environmental Quality (CEQ) has been central in strengthening procedures and guidelines in the USA since the National Environmental Protection Act 1969 (NEPA) was enacted. Public scoping of all federal agency projects requiring an EIA is the norm. Publicity in various media is combined with invitations for comments or suggestions, which can be submitted via the post, at various public open houses or public meetings, and via e-mail. Sri Lanka. The scoping process here involves formal and informal meetings with people who may be affected by the proposed project either directly or indirectly, or who may have special knowledge of the project area and its environment (Tille Kerante, L.S.G., 2003). Australia. The Environment Protection and Bio Diversity Conservation Act 1999 (EPBC Act) establishes a Commonwealth process for environmental assessment and approval of proposed actions likely to have a significant impact. This is the culmination of over 25 years of EIA in Australia.

Five options are listed under Section 8 of the EPBC Act (Environment Australia, 2003). The first three, preliminary documentation, public environment report (PER) and environmental impact statement involve the following steps: proponent prepares and publishes draft document; public comment period; proponent finalises document incorporating comments; Secretary of Environment Australia reports to the Commonwealth Environment Minister who makes a decision. All documents prepared by the proponent will follow guidelines (terms of reference) laid down by Environment Australia, and formulated by consultation with the relevant stakeholders. Where the impacts are likely to be high, and prior to preparation of the final EIS, or where greater public involvement is to be ensured, a Public Inquiry is held. All information leading up to the inquiry, including the reasons for having same, are made available.

The hearings are all held in public and all submissions made available to the wider public. Quebec, Canada. (Australia EIA Network, 1996 p.25) Established in 1978, the Bureau d'Audiences Publiques Sur L'Environnement (BAPE) is a quasi-judicial advisory body with a strong record of public participation. The Minister on request from any person, group or municipality may instruct BAPE to hold; a fact finding enquiry; carry out a combined enquiry and mediation; or undertake public hearings. Its role is to enquire into any question, relating to the quality of the environment, that has been submitted to it by the Minister of the Environment and report back on its findings. It is not like An Bord Pleanala, which is an appeals board only. B.A.P.E. once instructed, plays a proactive role.

A BAPE Commissioner meets with the parties and mediates to seek a joint solution. Failing this, two to five Commissioners serve as an independent review board. Initially, the public has 45 days to review an EIA and request mediation or a hearing. BAPE must report to the Minister, within 120 days of receiving a request for a public hearing or 60 days from receiving a request for an inquiry or inquiry with mediation. Success rates of mediation by BAPE are 75%. The option for a further period for a hearing helps circumvent the feeling of negotiating under time pressure. At the same time, proponents know that these timescales are absolute. The mediation process provides an effective and efficient means of resolving conflict with a 75% success rate in 1978. Institutional arrangements for EIA Review (Australian EIA Network (1996). Permanent Commissioners in Holland (Netherlands Commission for EIA, 2002) and Italy review the quality of the EIS. The Dutch Commission for EIA remains outside political decision making, it's role is to give an expert (selecting from a panel of 400) opinion on the quality and correctness of an EIS. The Italian Commission by contrast can make a judgment on the environmental compatibility of a project, as well as the quality of its EIS. In Canada, for certain major projects (50 in the period 1975-1995), ad hoc review panels, or inquiry bodies are used. A panel's major responsibility is to establish terms of reference (TORs) for an EIS and to conduct a public review of its suitability. The Parliamentary Commissioner on the Environment in New Zealand can also use panel reviews.

Ireland, Planning Major Roads. Although the Roads Act, 1993 as amended by the Planning and Development Act, 2000 only specify public consultations in the case of compulsory purchase orders (CPO's) or post EIS publication, the NRA in its 'National Roads Project Management Guidelines' sees public consultation as a vitally important and integral part of the various phases of major project planning. The first round of public consultation occurs in the preliminary planning stages as part of the feasibility/constraints study. Following the publication of the Consultants Report, which is based on the public submission and the consultants report, the possible route options and their general impacts are identified and the public consulted again. This leads to the route selection, and only then, is the EIS drafted. (NRA, 2000).

Research Into EIA. The EC Joint Research Centre (ISPRA) is at present conducting a study on behalf of the EC Environment DG into EIA and SEA research. One of the six areas for research relating to EIA's is public participation (Europa, 2003). A more general research 'Programme on Environmental Decision Making' (2001-2006) (E.S.R.C. 2001) aims to investigate and elucidate a set of criteria for better environmental decision. Further examples of good practice in public involvement in EIA and other areas of Environmental governance can be accessed through the good practices suite of the United Nations web-site: Virtual Conference for Integrating Environmental Considerations into Economic Policy Making Processes (ESCAP).

STRATEGIC ENVIRONMENTAL ASSESSMENT (SEA)

"SEA is a systematic process for evaluating the environmental consequences of proposed policy, plan or programme initiatives, in order to ensure they are fully included and appropriately addressed at the earliest appropriate stage of decision making on part with economic and social considerations". (Sadler and Verheem 1996). In their principles for SEA, Dalal-Clayton and Sadlier(Donnelly, A. et al, 1998, pp 31-42) call for transparency and openness, including early consultation, availability of comprehensible documents and explanation of how all decisions are arrived at.

The development of a pro-active SEA process is a subject too big for this study, but is one in which public participation should play a part as large as that proposed for the EIA process. The Dutch systems for EIA and SEA are identical in structure, and in both public involvement at the scoping phase is required. (Netherlands Commission for EIA 2002). By contrast, the new European Directive leaves consultation until much later in the process. 'Directive 2001/42/EC (The SEA Directive) (117) states that prior to the adoption of a plan or programme, or its submission to the legislative process, the competent authority must carry out an environmental report.

The draft plan or programme and the environmental report must be made available to the relevant authorities and to the public, who will be able to express their views on same, prior to it's adoption or submission to the legislative process. The relevant authority, following where necessary trans-boundary consultations, will inform all the parties concerned which have been consulted of its decision. With the decision will be a statement summarizing amongst others the results of the consultations, and the reasons for choosing the final plan or programme.

THE SURVEY

The purpose of the survey was to make an assessment of the present state of public participation in EDM in Ireland. An attempt as made to look at:

o Attitudes of the main stakeholders.

- o Their experiences with public participation.
- o Present in-house capacity of regulators and industry to enable public participation.
- o In-house capacity building measures.
- o Capacity building in the wider community.

Methodology

The questionnaire was based on the results of the literature survey, personal experience and feedback from trial questionnaires. The survey was conducted by e-mail, except the industry section, which due to confidentiality issues had to be posted by the college administrator. Where appropriate, the same questions were asked of all stakeholders, but specific questions relevant to each group were also included.

The public questionnaires included a preliminary set of open-ended questions. These were designed to establish the person's particular experiences and involvement in dealing with a particular EDM process.

Copies of all the questionnaires can be seen in Appendix III.

The main groups of stakeholders surveyed were as follows.

The Public.

Industry.

Local Authority, Environment Sections.

Environmental Awareness Offices (EAOs).

Department of the Environment and Local Government.

Other relevant departments and agencies.

All respondents were guaranteed confidentiality.

The Public-20 completed questionnaires.

Notice of the survey was given to all the ENGOs on the ENFO address list, and information about the research was given in the 'Horizons' column of the Irish Times (21/12/02 and 22/3/03). All the respondents had been involved in some form of EDM process, either individually or more usually in groups of from 4 to 2000 members.

The areas of EDM covered by the respondents included.

- o Urban, Rural, Waste, Landscape, Road and County Development Planning.
- o The EU Habitats Directive.
- o Integrated Coastal Zone Management (ICZM).
- o High Voltage Power lines.
- o Compliance with EC Directives.
- o Aquaculture.
- o Incineration.
- o Recycling.

Industry-15 Completed Questionnaires.

Postal contact was made with 40 graduates of the H.Dip in Environmental Protection. The responses came mainly from Environmental Engineers and Environmental Safety Officers, together with one consultant and one company CEO.

The industries involved included:

- Wind energy Distillation
- Pharmacies Biological
- Metallurgy Galvanizing
- Chemical Coating Turbines
- Aircraft Man-made fibres

CAPACITY FOR PUBLIC PARTICIPATION

From the responses to B6 and B7 it is clear that 60% of the industry do not have formal lines of communication with the wider community or help with comprehension of technical documents, and are generally passive in this regard.

By contrast, Local Authority (LA) staff do see a strong role for themselves in providing assistance, and in understanding legislation and technical documents, even though only 58% have clearly laid down procedures for requests for information. Even then the procedures are generally only in relation to requests under the Freedom of Information Act 1997. Only 27% of LA respondents have a freephone call-line, compared to 40% in industry. LA staff assess their own capacity, to give good access to information on the environment, very highly. (B11) Looking at B8 and B9 it is clear that the council staff are inclined towards the more traditional, generally passive methods of information giving but with a strong reliance on website development. EAOs by contrast and by the nature of their job, are more pro-active and use a wider range of tools. Many local authorities are still developing their websites, and so the availability of information on the different LA websites varies dramatically in quality and quantity, as can be seen from responses to B10 and B12.

B14 responses showed a generally low public access skills base, with not surprisingly, EAOs being better trained overall.

The sources of information were used by all the stakeholders was the subject of B16. The Public generally used independent sources, although surprisingly few used the excellent resources of ENFO. The use of the local authority, as a source, received a rating of 20% of the public total. Council staff also made little use of ENFO, sourcing most of their information with the EPA and the DOELG with some use of consultants. By contrast then ENFO was the first choice of EAOs, with the EPA and DOELG second.

Reflected here may be the lack of trust felt by the public in the regulatory authorities. It should be noted that the DOELG was not offered to the public stakeholders as a choice. The Local Authority is a necessary source of some information for most EDM processes and so received a larger number of individual public responses than the other sources, but its average score was 2.4 compared with 4.8 for "other" and 3.8 for "experts".

To summarise, the capacity to provide good access to public participation is at best patchy, despite the response by local authorities, to B11, asserting c.90% capacity to give good access to information.

Whilst the inter-net is an invaluable public access tool for public bodies and industries, the existence of the 'digital divide' is not mentioned by anyone. Most of the websites are still at an early stage of development, and very few are interactive. Yet the inter-net seems to be the main focus for the council respondents in particular.

Capacity Building

The response of the Industry respondents seemed to be generally aimed at increased transparency and was marked by a desire to reach out to the community. Both of the Local Authority respondent groups were looking for increased resources, staff training and improved websites. Increased public access to Strategic Policy Committees and regular meetings with key stakeholders were both good capacity building suggestions. A minority in both EAO and Council respondents felt no change was necessary to improve access to decision-making, with two EAOs suggesting that access to county councilors was sufficient. Question B18. What capacity building measures for public involvement in EDM would you like to see developed in your organization? (Please select and rank your top three)

a) staff training on legislation and technical details.

b) increased ease of access to information for staff.

c) staff training in public access skills.

d) increased advertising budgets.

e) appointment of a public information officer.

f) training in web-site development.

g) other/s (specify).

All the stakeholder groups seem to focus on skills training and access to information for their own staff. Industry looks for staff training in legislation and technical details, as do EAOs, whereas the main priority for council staff is training in participation skills and greater ease of access to information for staff members.

EAOs in particular want increased budgets for advertising, which is a key part of their work.

Please select the five most useful, and rank them with no.1 being your first choice etc. Percentages shown represent the cumulative percentage scores from each of the stakeholder groups.

Cumulative

% Scores.

A community based, independent environmental information broker. 37%

An environmental ombudsman. 58%

A one-stop information hotline with guaranteed call back. 29%

Technical assistance grants. 28%

Science shops*. 13%

Citizen training in how to use the EDM processes. 33%

Collaborative participation where all stakeholders (including developers and concerned citizens) share information and decision-making. 60%

Increased availability of information in various formats. 49%

Grants to community environment groups for educational purposes. 32%

Improved access to documents. 26%

An independent 'expert' panel for review of technical information. 35%

Total: 400%

*Science shops do community-based research where citizens set the agenda. University or college based, the talents of the students, supported by faculty advisers are used to answer questions posed.

This question is based on a list of measures, designed to build the capacity of the public for participation in the activities of the US EPA. (Breggin, L. and Hallman, H.)

Discussion of Question B19.

The results for this question are based on the cumulative % scores for each stakeholder group.

This was done to avoid distortion of the bar diagram due to the different sample sizes in each stakeholder group. Caution should be used in interpreting these results. (The percentages given here are of the score within each stakeholders group). Collaborative participation' was the overall first choice followed by the 'ombudsman' and then 'increased information'.

The Publics first choice was 23% for collaborative participation, with the environmental broker and the ombudsman being equal second choice at 16%.

Industry gave 24% of its overall score to the environmental ombudsman, 14% to increased information and 10% to the broker.

Council respondents gave their highest score (14%) to a "one-stop hot-line" with collaboration

13% and increased information 12%. They also gave 11% to each of 'technical grant assistance', 'improved document access' and 'expert panel'.

EAOs scored 17% for 'increased information availability', 15% for ' collaboration' and 11% each for 'environmental broker' and 'ombudsman'.

Combining the percentage scores for the 'broker', 'ombudsman', and 'expert panel', captures 44% of the Industry score, 41% of the Public score, 27% of the EAO score, and only 18% of the Council score.

In summary then, from the above, it is clear that there is a consensus among the stakeholders for working together in EDM processes. At the same time, there is a recognition that some issues need arbitration, as well as independent assessment of disputed 'facts'.

As it is usually proponents of projects and the public that are in dispute, it is not surprising to find them opting for arbitration whereas the council officials are looking more at the practicalities of file access, hot-lines and the availability of information, including technical information. The EAOs, not surprisingly since it is part of their job specification, scored highly for increased availability of information.

Industry. It was felt by two of the respondents that the social and political climate was not mature enough for public participation (PP). Others called for openness with a focus on working with small groups.

Councils. Generally, suggestions given were of a positive nature. It was suggested that the public should be brought in as early as possible to EDM processes. It was also suggested that PP be made an integral part of all EDM processes. Even if processes are slowed down, you cannot, according to one respondent, consult too much.

EAOs. PP in EDM requires openness and patience, and is best done face to face. The public should be made aware of the constraints on the council. People are only interested in contentious issues. Involve other community bodies.

RBM. Have sufficient resources at the outset to conduct your participatory process. Make sure, that the public are aware of the limits of the process, at the beginning. Make use of trained facilitators.

DOELG. Make people aware of their right to access information on the environment. Discussion.

Most of the responses were positive towards PP, though the DOELG talked only of access to information. Those that pointed to the political immaturity, and the interest in only contentious issues, pose the question in the author's head "how is this situation going to change?"

Summary and Discussion of Results

Given the inevitable drive to build a sustainable society in which the economic sociologic and environmental aspects of development are all considered, the importance of including all the stakeholders in environmental decision-making (EDM) becomes increasingly apparent. All actions for change have consequences, and the moves to greater public participation have their costs as well as benefits.

It is necessary for all people to decide whether these costs are greater or lesser than the costs of no change. In order to make that choice it is essential to have as much information as possible. This study attempted to draw together some of the arguments for and against that change. Making the assumption that some changes are coming already, as a result of the Aarhus Convention and its offspring, this study has examined ways of making public participation in EDM effective.

In understanding any change, it is necessary to know where you are starting from, and the possible routes that might be traveled to get there. To this end a survey was made of the major stakeholders in EDM processes, industry, the regulators, and the public, in order to see the starting point, as well as some of the possible ways forward.

In the time-scale of this treatise, it was not possible to get statistically representative samples of either industry or public opinion, although the local authority sample was sizeable. 35% of environment departments, 47% of EAO's covered in total 65% of Local Authorities. The methods by which the public and industry groups were contacted were very specific, so this and their self-selection in responding should be taken into account in interpreting the results. Every ENGO on the ENFO list was surveyed, but only one member of one organization responded, despite each receiving two reminders. Several said they did not have the time.

The authors impression was that they mostly saw public participation as an irrelevance to them. The IFA (Irish Farmer's Association), I.B.E.C. (Irish Business and Employers Confederation) and all the government departments with an obvious environmental remit were also surveyed to no avail, apart from the DOELG, who did respond.

From this the conclusion could be drawn that either these bodies are not interested in the survey, or they are not interested in public participation, or they see participatory practices as a dilution of their power. This is an area for future research. In order to get a reasonable return on the survey, it was necessary to send out the questionnaire long before the literature survey was completed. This meant that some questions that should have been asked were not, and others were asked badly. An example of the latter would be Question B.1.-Should the public be excluded from any particular area of EDM? The strong response against exclusion does not, however, indicate whether they think the public should be included. With regard to the former, it would have been helpful to find out why each group thought public participation should entail.

Taking the above reservations into account, and in the context of the literature survey, a number of indicators for future action can be taken from the results of the questionnaire survey, as will be seen. The questionnaire looked in turn at the experiences and opinions of the stakeholders, as well as their capacity to deliver public access to information and EDM. Opinions on capacity building measures were also sought.

Putting together the experiences of the different stakeholders, it fast becomes apparent that they have been very different.

On the one hand, we had the public stakeholders, who clearly felt that they have been largely ignored despite their best attempts at constructive engagement with the regulatory authorities. Their commitment cannot be denied in both time and money, and yet they seem to have met, in general, with unhelpful and sometimes obstructive treatment by public servants at local and national level. On the other hand we heard from those people working in Local Authorities and the DOELG that they were confident in their ability to provide good information to the public, and that although it might slow EDM processes down, they had the majority opinion that public participation had many positive attributes. They also felt that their staff were enthusiastic (c65%) though apprehensive about public involvement in EDM. Industry respondents were even more enthusiastic for public involvement, though they did not have the same strength of belief in its ability to reduce conflict or improve project design.

Clearly something is wrong here. Assuming that all the respondents are genuine in their responses, and that it would appear that they would all wish to operate in a more open way, with the protection of the environment as an end point, there is clearly either a communications breakdown, or else the structures and laws set up to protect the environment are unable to deliver public participation.

It seems like a mixture of the two. In order to communicate well, and under time pressure, it is necessary to speak the same language and to know the rules of engagement. It seems that one third of the public did not understand the EDM processes they were entering into, and only two thirds understood at the end, with very few (27%) getting any assistance along the way. And yet the public servants seem more than willing some (c80%). They were also willing to help with technical matters (c75%).

Independent expertise was hard got by the public and expensive, and yet in situations where the proponent of the project or plan was a government body, they saw themselves as paying for expertise on both sides of the debate.

What happens when a member of the public requests information? Only 58% of Local Authorities said they had strictly laid down procedures and only 40% of industry. Where guaranteed call back hotlines operate (only 25% of Local Authorities) response times were from twenty minutes to 72 hours.

It would be useful here to know how the front-of-house people in Local Authorities feel about their ability to deal with these questions. Do they have the training, back-up and staff numbers enable them to give the service they would like. These questions, unfortunately, were not put. However, in looking at questions relating to capacity, the following becomes clear. For enabling information delivery, the web-site is a powerful tool but limited in its accessibility, and yet it seems to be the focus of most of the council respondents. At the same time, despite some shining stars, environmental information on Local Authority web-sites, is still fairly restricted, as in that on the E.P.A. and DOELG web-sites. However, they are all in the process of improving, and in the case of the latter two, undergoing radical overhauls. Links from Local Authority web110 sites to other sources of information ranged from 50% for local bye-laws, down to 29% for European Directives.

The author is not aware of any research into who accesses the existing sites, or who knows that they exist, and yet they seem to be the main focus for government resources, and local government officials, in terms of improving accessibility.

Whilst EAO's, whose job is to educate and interact with the public on a daily basis, rate their public access skills highly, they are not the ones on the front line in EDM processes. The percentages of Industry respondents in particular and Council respondents that replied positively to having public access skills was low in general, except in media skills, and for the council officials with regard to public meetings.

It is not to be expected that these two groups would be highly trained in these areas, it is not their first priority, but these skills, particularly the facilitators skills needed to work with small groups and in workshops, should be available, in house, in every government agency or Local Authority dealing with the public in EDM.

These skills would also be useful in other aspects of their internal and external business, e.g. Partnership and the development of E.M.A.S. Generally though, when asked about proactive methods of giving information and receiving feedback, industry was very passive in its approach. Council officials also tended to be passive but showed a greater willingness to engage with the public. As mentioned earlier, the web-site seemed to be the tool of choice for council officials.

The EAO's and the DOELG were both very proactive in information giving and feedback but the EAO's as educators showed the widest range of skills.

Access to outside sources of information is another area of capacity that needs examination. Question B.16. showed council officials, who deal largely with the technical/decision making aspects of EDM, opting for sources of information that enable them in these tasks. Similarly EAO's, who used the educational resources of ENFO also used the DOELG and EPA. By contrast, the public largely used independent sources which got very high average scores of 4.8 111 for 'other' and 3.8 for 'experts' compared with only 2.4 (out of 5) for the Local Authority, even though 75% of the respondents were involved in EDM processes to do with Local Authorities and planning. None of the public respondents were from Dublin, which might explain the low use of ENFO (8%). Those that did, gave it a score of 3.2 on average. The author would score ENFO much higher as a resource, from his own experience, and would recommend that ENFO be resourced such that it can make its presence felt in every region of the country.

The use of independent sources of information by the public would tie in with Question A.11 where only one third of the 75% of respondents dealing with local government planning issues had increased trust at the processes end, and a lower figure dealing with waste permitting (22%). Trust in the EPA was zero and in An Bord Pleanala 43%.

The only area of actual EDM discussed in the questionnaire was the EIA process. Clearly the public respondents that had had experience of EIS's found them of very poor quality indeed. Alongside this, we have the Industry respondents, who were 73% enthusiastic about public involvement in EDM, giving only 57% support for greater public involvement in the EIA process, with 53% agreeing with public involvement in the 'comparison of alternatives'.

So this is the base line, we had a largely disgruntled public, an enthusiastic but under-skilled and under-resourced public service, and an industry response that was generally lukewarm. How did these same respondents see the way forward, what capacity building did they see as useful? When asked what changes in their organization would improve public access to environmental information and EDM, Industry respondents look to increase the transparency of their enterprises, and showed a desire to reach out into the community.

Local Authority and EAO respondents both looked for increased resources, staff training and improved web-sites. Regular meetings with key stakeholders in the community, coupled with increased public access to Strategic Policy Committees were both creative suggestions.

When asked to make choices about specific capacity building measures, Council respondents opted for staff training in public access skills and increased access to information for their staff; whilst the main choices for industry were staff training in legislation and technical details alongside training in public access skills. EAO's looked for advertising budgets to assist with their education programmes.

The penultimate question, B.19, dealt with capacity building measures for the wider community,.

The measures receiving the highest cumulative scores were; 'collaborative participation where all stakeholders share information and decision-making' at 60%, and 'an environmental ombudsman' at 58%.

Apart from this, Industry and the Public both scored very highly for the three interventionist measures of; an expert panel to resolve disputes over 'facts; a community based independent environmental information broker, to give that expertise sought by the public; and an ombudsman to act as a final arbiter if all else fails.

The final question B.20, asked of all, but the public, looked for advice to others, based on the experiences of the respondents. Most of the responses were positive to PP, suggesting a more personal approach to the EDM processes with PP incorporated into the EDM process at the earliest possible stage.

Two responses from Industry felt the politico/sociological climate in Ireland was not mature enough. Perhaps by reaching out into the communities, of which they are but one part, they might be surprised. There is a great wealth of detail in the responses of all the stakeholders, which can be seen in Appendix II and which in its variety emphasizes the fact that, as in all human activity, nothing is black or white.

CONCLUSIONS

It is clear that the present situation with legislation, that largely restricts public participation to 'report and comment' activities, is likely to leave the public stakeholders with feelings of frustration and alienation. It is also clear, that without the necessary changes in practice, and increases in resources for skills training and more personnel, public servants cannot deliver effective access to the public. The lack of trust shown by the public in the regulatory authorities, put together with the apprehensions of the regulators and industry, about public involvement in EDM, will require a great deal of work to change. Open and inclusive debate, amongst all stakeholders, about what changes are needed in their inter-relationships and in the distribution of decision-making power are essential.

The Directives coming from Europe and emanating from the Aarhus Convention should be seen as only the minimum standards, and the spirit of Principle 10 brought in when they are enacted into Irish law. Indeed, the drafting of new legislation should itself be open to public consultation. Greater public awareness of their rights to be involved in EDM processes, and greater knowledge of how to be involved would, it was agreed, lead to the end-game of greater environmental protection. The more the public are involved in EDM, the better educated they will become on environmental issues, and hopefully then the safer the environment will be from thoughtless plundering and degradation of natural resources.

The way forward then lies in; greater openness, a willingness, on the part of regulators and industry, to reach out pro-actively into their communities, a detailed assessment of the capacity needed to deliver participation, and a commitment on behalf of all stakeholders to work together for their mutual benefit, and a sustainable future for all life in this Biosphere of which we are a small but potentially very destructive part.

RECOMMENDATIONS

- That Ireland, as a signatory, actively seeks to carry the spirit of Principle 10 into all areas of EDM.

- That Irish Governments, in drafting legislation on the environment, proactively consult all stakeholders beforehand.
- That the Irish Government finance an assessment of Principle 10 implementation, as agreed under Partnership for Principle 10, using an independent assessment method such as the Access Initiative. In this way, capacity building will have a base-line to focus on.
- That regulators proactively seek out stakeholders and encourage their involvement in EDM well before the final decision is made e.g. calling door to door in the neighbourhood of a proposed development.
- That timescales for submissions about EDM should be more flexible, allowing more time for the consideration of complex issues.
- That funding be made available, to enable public participation, in the form of grants for legal and technical assistance.
- That the public be involved as early as possible in the EIA process. Preferably from the screening phase onwards, sharing in the information flow that leads to the EIS.
- That reasons be publicly given for the final decisions in all EDM including the decisions of An Bord Pleanala, and that responses to submissions are included with same.
- That the role of ENFO should be expanded both geographically and in terms of its service provision. This latter to include an independent environmental information broker, with access to a panel of independent experts.

5

ENVIRONMENTAL POLICY AND LAW

INTRODUCTION

The Philips Global Environmental Policy is a core operating element for the operations of the entire Philips organization. Since 1970 the Board of Management of Philips Electronics has formulated several guidelines for environmental performance. A general environmental policy was issued in 1987 and updated in 1991. The Philips Global Environmental Policy is a living document – it is part of the fabric of the organization and is reviewed and updated to meet the needs of our stakeholder and the environment.

PHILOSOPHY

Philips believes that sustainable development is one of the most challenging issues facing the world. The company is committed to continuously exploring solutions to successfully balance economy and ecology.

Basic principles

Sustainable development – finding the optimal balance between ecological impact and economical growth. Prevention is better than cure – avoiding problems at the óutset. Life cycle approach – from raw materials, to manufacturing, to use and disposal. Cooperation – with governmental and non-governmental organizations.

POLICY

- Philips establishes technically and economically viable

objectives to optimize the environmental performance of the organization's products, services and activities.

- Product development objectives include evaluating the environmental impact over the total life cycle of a product, taking steps toward more efficient use of materials, including packaging; reducing or eliminating hazardous substances; reducing energy consumption; and contributing to improving recycling and disposal. - Manufacturing objectives include environmentally related aspects of Philips activities such as emissions into air and water; use of energy and water; and waste disposal.
- Philips establishes and maintains environmental management systems and audits them by means of a systematic, documented verification process to ensure continuous improvement.
- Philips is committed to complying with all applicable laws and regulations, and will promote international harmonization of applicable laws and regulations, and is prepared to enter into voluntary agreements.
- Philips communicates its environmental policy to employees and other stakeholders, and publishes its results in environmental reports.
- Philips educates its employees to work within its environmental policy.

GENERAL CLASS POLICIES

1. *Attendance Policy:* As graduate students, you are expected to attend every session, especially given the emphasis on class discussion. However, if you find that you have to miss a class, advance notice to the instructor would be appreciated. Please note that missing more than one or two classes will have a serious impact on the student=s participation grade.
2. Make-up exams will be given only during the scheduled final exam period. Make-ups are discouraged and only the best of reasons will permit make-ups. If you miss the mid-term,

you must inform the instructor of the missed test within a week of the scheduled test and remind the instructor of the need for the make-up during the last week of classes.

3. Class materials (e.g., syllabus, study guides, return of exams) can be picked up only during scheduled class periods. None of these materials will be given outside of class.

4. Discussion, participation and questions from the class are especially encouraged. I will also be happy to talk with you during my office hours or by appointment. If you have any problems (for example, shyness, a handicap or illness), please come and talk with me. Missouri State University makes reasonable accommodations for students with disabilities. Please see the instructor if you need reasonable accommodation.
5. *Disability Accommodation Policy:* To request academic accommodations for a disability, contact Disability Services (DS), Plaster Student Union, Suite 405, (417) 836-4192 (voice); (417) 836-6792 (TTY) http://www.smsu.edu/disability. Students are required to provide documentation of disability to DS prior to receiving accommodations. DS refers some types of accommodation requests to the Learning Diagnostic Clinic (LDC). The LDC also provides diagnostic testing for learning and psychological disabilities. A fee is charged for testing. For information, contact the LDC (417) 836-4787; http://www.smsu.edu/contrib/ldc/.
6. *Academic Integrity:* Missouri State University is a community of scholars committed to developing educated persons who accept the responsibility to practice personal and academic integrity. You are responsible for knowing and following Missouri State University's student honor code, Student Academic Integrity Policies and Procedures, available at http://www.smsu.edu/acadaff/AcademicIntegrity.html and also available at the Reserves Desk in Meyer Library. Any student participating in any form of academic dishonesty will be

subject to sanctions as described in this policy.Cheating on exams and/or term paper (i.e., plagiarism) is obviously discouraged and will be treated appropriately. The usual penalty for cheating on exam or plagiarism of a paper is a failing grade for that assignment.]

7 *Nondiscrimination Policy:* Missouri State University is an equal opportunity/affirmative action institution, and maintains a grievance procedure available to any person who believes he or she has been discriminated against. At all times, it is your right to address inquiries or concerns about possible discrimination to Jana Estergard, Equal Opportunity Officer, Siceluff Hall 296, (417) 836-4252. Other types of concerns (i.e., concerns of an academic nature) should be discussed directly with your instructor and can also be brought to the attention of your instructor's Department Head.

8. *Cell Phone Policy:* As a member of the learning community, each student has a responsibility to other students who are members of the community. When cell phones or pagers ring and students respond in class or leave class to respond, it disrupts the class. Therefore, the Office of Academic Affairs prohibits the use by students of cell phones, pagers, or similar communication devices during scheduled classes. All such devices must be turned off or put in a silent mode and cannot be taken out during class. At the discretion of the instructor, exception to this policy is possible in special circumstances.

9. *Disclaimer:* The instructor may change any of the course requirements as needed if notice is given at least one week prior to the scheduled event. The dates for the paper and the final are firm and will not be changed.

NEPA applies to certain proposed actions of the Federal Government. NEPA does not apply to the independent actions of States or private property owners. It has no applicability to a private property owner's use or development of his/her property rights, nor the development of a State's FLP. It could apply to Federal agency actions undertaken on private property if the U.S. acquired a right to permit or deny certain land uses and then proposed to exercise that right, but in such an instance it would be the U.S. that would be required to satisfy NEPA requirements, not the private owner. It should be known that:

1. A Programmatic Environmental Assessment and a Finding of No Significant Impact was completed for the national FLP and signed by the Chief of the Forest Service.
2. Under the Federal acquisition option, the FS NEPA regulations (Forest Service's Environmental Policy and Procedures Handbook 1909.15-92.1, effective 9/21/92), the acquisition of an individual Forest Legacy tract and/or easement may be categorically excluded from the preparation of an Environmental Impact Statement or an Environmental Assessment unless scoping indicates extraordinary circumstances exist.

COORDINATION WITH STATE FORESTRY AGENCIES

Whereas most State lead agencies are State Forestry agencies, and the CFAA establishes a broad cooperative relationship between the FS and State Foresters, the FS shall appoint a representative to coordinate with the Working Lands Conservation Committee of the NASF (or its successor) regarding the FLP. Periodically, the Director of Cooperative Forestry, and the appointed FS representative shall meet with the NASF Working Lands Conservation Committee to assess program operations, accomplishments, and policies. In States where the State Forestry agency is not the designated State lead agency for the FLP, a coordinating mechanism shall be instituted between the State lead agency, the State Forester, and the SFSCC.

Assessment of Need (AON) and Identification of Forest Legacy Areas (FLA)

A State or a federally recognized Indian tribe conducts an AON, in cooperation with the SFSCC, to document their need for inclusion in the FLP, through an evaluation of current forests, forest uses, and the trends and forces causing conversion to nonforest uses. Federally recognized Indian Tribes must cooperate with the SFSCC when conducting an AON for nontrust allotments lands. The AON is intended to define the Eligibility Criteria to be used in the identification of important forest areas to be proposed as an FLA; identify and delineate the boundaries of forest areas meeting the Eligibility Criteria for designation as an FLAs; determine through analysis what defines "threatened" and "environmentally important forests"; and outline the State's project evaluation and prioritization procedures. The AON must be developed in consultation with SFSCC and approved by the State lead agency.

State lead agencies may utilize the services of land trusts or other entities in preparing the assessment. Information from existing sources may be used to prepare the AON, instead of initiating new studies that would duplicate existing data. Examples of appropriate sources include State Forest Resources Plans, State Comprehensive Outdoor Recreation Plans, growth management studies, State cultural site inventories, inventories of threatened and endangered species, and other State, regional or local plans, studies or reports. The AON shall include relevant information about both public and private lands, address the issue of how best to maintain the integrity of forestlands for future generations, and address pertinent issues as identified by the State.

At a minimum, the AON must address the following as they relate to the purpose of the FLP:

1. Forest resources including:
   - o Aesthetic and scenic values;
   - o Fish and wildlife habitat;
   - o Minerals resource potential;
   - o Public recreation opportunities;
   - o Soil productivity;
   - o Forest products and timber management opportunities;
   - o Watershed values including water quality protection;
2. The present and future threat of conversion of forest areas to nonforest uses. States are responsible for defining the conversion threat(s);

3. Historic uses of forest areas, and trends and projected future uses of forest resources;
4. C urrent ownership patterns and size of tracts, and trends and projected future ownership patterns;
5. Cultural resources that can be effectively protected;
6. Outstanding geological features;
7. Threatened and endangered species;
8. Other ecological values;
9. Public recreational opportunities;
10. Protected land in the State, to the extent practical, including Federal, State, and municipal lands and land trust organizations lands;
11. Issues identified by the SFSCC and in the public involvement process.

Using the above information the AON shall include the following:

1. Identification of applicable Eligibility Criteria;
2. Identification of specific FLA(s) for designation;
3. Specific goals and objectives to be accomplished by the FLP;
4. Process to be used by the State lead agency to evaluate and prioritize projects to be considered for inclusion in the FLP.

The project evaluation and prioritization process outlined in the AON should reflect the direction set forth in the CFAA to give priority to lands which can be effectively protected and managed, and which have important scenic or recreational values, riparian areas, fish and wildlife values including threatened and endangered species, or other ecological values. Traditional forest uses such as forest management activities, including timber management, and outdoor recreation opportunities are considered consistent with purposes of the FLP and are encouraged on FLP tracts when consistent with the State's AON and the conservation purposes for FLP tract acquisition. The prioritization process should implement a strategy that enhances existing protected forestlands or local and State conservation strategies as outlined in the AON.

The composition of the SFSCC is defined in Section 19(b) of the CFAA (16 U.S.C. 2113). States are encouraged to broaden this composition to include interests appropriate to benefit the FLP. This committee cooperates with the State lead agency in the preparation of the AON, identification of FLA Eligibility Criteria, the identification of proposed FLAs from which lands may be entered into the FLP, and recommendation of priority lands to be considered for enrollment in the Program.

Public participation and involvement in the AON preparation is a State responsibility. In the absence of established State procedures, NEPA may serve as an appropriate model for public involvement. The State lead agency will solicit involvement and comments on the AON from the public including State and local governments. The goals of public involvement include hearing concerns and views from interested and affected individuals and organizations, receiving new information, identifying and clarifying issues.

Based on the State-wide AON, the State lead agency, in consultation with the SFSCC, identifies specific geographic FLAs that meet the Eligibility Criteria, and recommends them to the FS for designation as of a FLA.

States are encouraged to cooperate in the identification of FLAs that cross State boundaries and to work together to coordinate acquisitions of lands or interests in lands that have complementary purposes. However, program implementation is undertaken by the individual States (State Grant Option) or by the FS (Federal Option).

The identification of proposed FLAs must include:

1. Location of each geographic area on a map and a written description of the proposed FLA boundary;
2. Summary of the analysis used to identify the FLA and its consistency with the Eligibility Criteria;
3. Identification of important environmental values, and how they will be protected and conserved;
4. The conservation goals or objectives in each FLA
5. List of public benefits that will be derived from establishing each FLA;

6. Identification of the governmental entity or entities that may hold lands or interests in lands (State grant option) or may be assigned management responsibilities for the lands and interests in lands enrolled in the program (Federal option); and
7. Documentation of the public involvement process and analysis of the issues raised.

ELIGIBILITY CRITERIA FOR ESTABLISHING FOREST LEGACY AREAS (FLAS)

The CFAA directs the Secretary to establish Eligibility Criteria for the designation of FLAs, in consultation with the SFSCC. These criteria should be based upon the FLP purpose to protect environmentally important forest areas that are threatened by conversion to nonforest and be further developed through the AON.

FLA boundaries must encompass forestlands with significant environmental and other resource-based values. Areas may also include nonforested areas such as farms and villages if they are an integral part of the landscape and are within logical boundaries. Since FLA boundaries may not correspond to property boundaries, tracts located partially within the geographically defined FLA are eligible for the FLP, upon approval of a boundary adjustment by the FS Region/Area/IITF.

Indian reservations and tribal lands may have important features on the forested landscape. Indian tribes and States are encouraged to collaborate and to consider only nontrust allotment lands for designation as, or inclusion within, a FLA. Other tribal lands are already protected through the trust relationship between the U.S. Department of the Interior and the tribe and are ineligible for the FLP.

States are responsible for determining what defines "threatened" and "environmentally important forest areas" in the State. However, environmentally important forest areas shall contain one or more of the following important public values, as defined by the States:

1. Timber and other forest commodities

2. Scenic resources;
3. Public recreation opportunities;
4. Riparian areas;
5. Fish and wildlife habitat;
6. Known threatened and endangered species;
7. Known cultural resources;
8. Other ecological values.

The FS, State or unit of State or local government may only acquire lands and interests in lands identified within a FLA under FLP authority on a willing seller/willing buyer basis.

AON and Amendment Approval

The State lead agency must submit the AON, including proposed FLAs and Eligibility Criteria, to the FS Region/Area/IITF. The FS Region/Area/IITF with input from the FS Washington Office reviews the AON and works with the State lead agency to complete the AON. Once finalized, the FS Washington Office forwards the AON to the Secretary for final approval. Final approval establishes the FLP for the State.

AONs shall be periodically reviewed (at least at 5-year intervals) by the State lead agency, the FS Region/Area/IITF, and the SFSCC to assess whether AON amendments or updates are necessary. The results of reviews will be documented by the State lead agency. AONs should be amended as needed.

The State lead agency may amend the AON to make significant changes or minor adjustments. Significant changes include modifications to their FLP, changes to the FLA Eligibility Criteria, or to add or delete a FLA.

These changes need to be made in consultation with the SFSCC and with public involvement. FLAs and project evaluation criteria shall be of a scale and detail to effectively focus delivery of the FLP.

Significant Amendments to an AON may address the following:

1. Issues associated with maintaining the integrity of forestland and the proposed FLA specifically.

2. Revision, if any, of the FLA Eligibility Criteria.
3. Changes in policies or conditions that have occurred since the previous AON;
4. The identification of proposed FLA(s) and conservation goals or objectives associated with that FLA.

The Chief of the Forest Service, or designee, provides final approval of the Amendment of an AON authored by a State lead agency after consultation with SFSCC. In addition, the State lead agency may complete minor AON amendments, such as FLA boundary adjustment or project prioritization process. These minor changes need to be coordinated with the SFSCC, and need review and approval by the appropriate FS Region/Area/IITF.

Multi-State Identification of Forest Legacy Areas

States are encouraged to cooperate in the identification of FLA that cross State or Tribal boundaries and to work together to coordinate acquisitions of lands or interests in lands that have complementary purposes. States may elect to jointly use an existing or new multi-State or regional entity to identify FLAs or develop FLP projects that cross State boundaries. The entity must be a government-established organization, whose jurisdiction encompasses all or portions of the land area of the FLA States involved. However, program implementation is undertaken by the individual States (State Grant Option) or by the FS (Federal Option). The entity conducting a multi-State identification of FLAs is responsible for:

- o Obtaining approval from the appropriate States or Indian tribes for FLAs within their boundaries,
- o Cooperating with appropriate SFSCCs,
- o Obtaining public comments on the identification of FLAs, and
- o Complying with all other requirements of these guidelines.

Project Selection Process

The FS will conduct a project selection process to arrive at a prioritized national project list for consideration in the President's budget for the upcoming fiscal year. The project selection process and calendar of due date milestones are developed in consultation with the State lead agencies and FS Region/Area/IITF and communicated by the FS Washington Office. The FS will ensure that national evaluation and prioritization criteria are communicated to the States and in a timely manner so that submitted projects adhere to strategic goals and objectives of FLP. Project selection steps are:

Step 1: Release Project Selection Calendar with Due Dates

The project selection process and calendar of due date milestones are developed in consultation with the States and FS Regions/Area/IITF and communicated by the FS Washington Office.

Step 2: State Project Prioritization and Submission

FLP project applications are accepted by the State lead agency as outlined in the State's AON. The SFSCC reviews and evaluates applications according to the criteria identified in the State's AON, authorizing statute, and other relevant direction and policy, and provides recommendations to the State lead agency. Projects approved and prioritized by the State lead agency are forwarded to the FS Region/Area/IITF for funding consideration. Only projects submitted through this process will be deemed eligible. Each State will prepare a list of projects and enter or update its list for submission to the FS via the Forest Legacy Information System or other means as requested.

Step 3: Forest Service Regional Review

FS Regions/Area/IITF will review submitted projects considering State priorities and national criteria. The purpose of this review is to improve project viability, facilitate the national project selection process and advance the strategic outcomes of the FLP. FS Regions/Area/IITF will submit projects to the FS Washington Office for funding consideration.

Step 4: National Review; Develop National Project List

The FS Washington Office will develop a prioritized national project list by convening a panel. There are 3 purposes of the panel;

(1) assure that all projects meet Congressional and Administration direction;

(2) assure that projects meet national program goals; and

(3) develop a National List of ranked projects.

The composition of the panel shall be developed annually in consultation between the State lead agencies and the FS, and will be representative of geographic regions. Project evaluation and ranking is based on the following national core criteria; project readiness will be considered as well as other evaluation considerations developed in consultation with State lead agencies and FS Regions/Area/IITF. The national core criteria are:

- o *Important:* The public benefits gained from the protection and management of the property including environmental values, and the economic and social aspects;
- o *Threatened:* Conversion to nonforest uses is likely or imminent and will result in a loss of forest values and public benefits; and
- o *Strategic:* Fits with a larger conservation plan, strategy, and initiative and enhances previous conservation investments.

States newly entering FLP will be given a "New-State start-up" preference for an initial FLP project. This is a placeholder for planning purposes and does not guarantee project funding. In order to receive the New State start-up project funds the State must have an approved AON and the project must meet national core criteria and the State's evaluation criteria and be submitted within the fiscal year that the placeholder is approved by Congress.

Step 5: Submit National FLP Project List to the Office of Management and Budget and to Congress

Each fiscal year, the FS Washington Office will submit a project list to the Office of Management and Budget for funding consideration in the President's budget. Once the President's budget has been completed, the FS Washington Office will notify the appropriate House and Senate Committees and Subcommittees of the recommended projects for the upcoming Fiscal Year.

PROGRAM FUND CATEGORIES

Forest Legacy funds are allocated to one of three categories: Project Funds, Administration Funds, and AON Preparation Funds. FLP funds may not be used for monitoring and enforcement.

Project Funds

Project funds are those used to directly purchase lands or interests in land joining the FLP. Project funds may be expended by the State lead agency or the FS, as applicable, to cover transaction costs, including but not limited to: appraisals and appraisal review, land surveys, closing costs, establishing baseline information, title work, purchase of title insurance, conservation easement drafting, and other real estate transaction expenses for those tracts. Project funds may also be expended to facilitate donations of land or interests in lands to a qualified and willing donee for FLP purposes, by paying for expenses directly related to the donation, including but not limited to, land surveys, conservation easement drafting, title work, and establishing baseline information. For an outright donation of a conservation easement or land, FLP program funds may not be used to pay for an appraisal. In the case of a partial donation of a conservation easement or land, an appraisal meeting Federal standards is required to determine the value of property. FLP funds may be used for appraisals on these partial donations. When Federal funds are used to purchase real property, including conservation easements, appraisal and acquisition work procedures must meet Federal standards.

Administration Funds

Administrative funds are the portion of funds used for day-to-day program management at all levels. Administration funds may be used for a variety of activities, including FLP program administration, personnel and overhead, and all activities identified as eligible uses of project funds to prepare projects and potential projects. Forest Legacy funds for administration shall be kept to a minimum. As a goal, all attempts should be made to keep administration funds under 15 percent of the total funds appropriated.

AON Preparation Funds

AON preparation funds may be made available to States to help defray the cost of preparing, or amending an AON.

Process for Allocating Funds to Forest Service Regions/ Area/IITF

Following passage of the annual appropriations bill, the FS Washington Office develops the Forest Legacy Program Direction and allocates funds to the FS Regions/Area/IITF for distribution. The allocation process differs for each fund category described below.

Allocation of Project Funds

Allocations to FS Regions/Area/IITF are based on the results of the national project selection process and the appropriations bill. Under the State grant option, FS Regions/Area/IITF will award grants to States for specific, identified projects.

Allocation of Administration Funds

The FS Washington Office distributes administration funds to FS Regions/Area/IITF. Each FS Region/Area/IITF in consultation with the States requests these funds to meet their needs and the needs of the participating states in their Region/Area/IITF. Administration funds are also used by the FS Washington Office to fund program management functions. Administration funds will be granted to States under the State grant option separately from project funds.

Allocation of AON Preparation Funds

The FS Washington Office distributes AON preparation funds to the States by way of FS Regions/Area/IITF. These funds are requested by FS Regions/Area/IITF to meet the needs of their States to develop new AONs or amendments.

Redirection and Reprogramming of Funds

Due to the nature of real estate transactions, FLP projects may change in scope, cost or fail completely. These changes can result in unspent or excess funds for some projects while others may need additional funding to bring them to completion.

In order to maximize the efficient and effective use of FLP project funding, the FS will either redirect or request reprogramming of funds. Redirection is a shift of funds from one congressionally approved project to one or more other congressionally approved project(s). Reprogramming is a shift of funds that exceeds an increase or decrease of 10% per project not to exceed $500,000 to an existing project, or shifting of any amount of funds to a project not previously approved by Congress.

Regional Redirection Process

FS Regions/Area/IITF may redirect up to an increase or decrease of 10% per project not to exceed $200,000 of project funds that are excess or unspent from one project to one or more other Congressionally approved project(s) within the FS Region/Area/IITF which is underfunded and where there is a substantiated need (e.g. loss of other funding sources, appraisal documenting increased cost, etc.) to bring the project to completion.

Project funds over $200,000, or those that cannot be redirected by the FS Region/Area/IITF, will be released for the national process. FS Regions/Area/IITF will notify the FS Washington Office before a redirection takes place and report these actions periodically. All funds from failed projects will be released for the national process.

National Redirection Process

The FS Washington Office, through consultation with FS Regions/Area/IITF, may redirect up to an increase or decrease of 10% per project not to exceed $500,000 of project funds that are excess or unspent from one project to one or more congressionally approved project(s) which is underfunded and where there is a substantiated need (e.g. loss of other funding sources, appraisal documenting increased cost, etc.) to bring the project to completion. In addition, when funds have not been spent or contractually obligated within two years of receipt of funds, they revert to the FS Washington Office via the appropriate FS Region/Area/IITF. The FS Washington Office will:

1. Assess the extent of unspent or returned funds on a periodic basis;
2. Facilitate selection and funding of underfunded projects not addressed by the regional process or between FS Regions/Area/IITF; and
3. Notify Appropriations Subcommittees of any redirection action taken by the FS FS Regions/Area/IITF or FS Washington Office.

National Reprogramming Process

The FS Washington Office, through consultation with FS Regions/Area/IITF, may request reprogramming by the Appropriations Subcommittees of unspent or returned funds to a project that requires more than an increase or decrease of 10% per project not to exceed $500,000 to complete. In addition, the FS Washington Office may request reprogramming by the Appropriations Subcommittees of unspent or returned funds to a project not previously approved by Congress. The FS Washington Office will:

1. Determine the funds available for reprogramming on a periodic basis.
2. Identify underfunded projects that cannot be addressed through the Regional Redirection Process and determine the priority for reprogramming.
3. Recommend reprogramming to fund projects from the National Project List next in sequence in priority ranking to the extent practicable.
4. Submit reprogramming requests to the Appropriations Subcommittees for approval.
5. Allocate funds to projects approved for reprogramming.

FLP Cost Share Requirements

The CFAA directs that, to the extent practicable, the maximum Federal contribution for total program costs may not exceed 75 percent. To assure program-wide cost share goals are met, each project budget must include a minimum nonfederal contribution of

25 percent. This nonfederal cost-share must meet Forest Legacy purposes. It may consist of:

(1) the value of land, or interest in land, dedicated to the FLP that is not paid for by the Federal government;

(2) nonfederal costs associated with program implementation; and

(3) other nonfederal costs associated with a grant or other agreement that meets FLP purpose.

The nonfederal cost-share must be documented, and in the case of a grant, must meet the timing, terms, and conditions of the grant. The cost-share can occur at any phase of the FLP including planning, developing future projects, acquisition, capital improvement, management, or administrative activities. When a grant is involved, the cost-share must occur within the life of a grant and meet all grant requirements. Federal requirements identify the grant period as beginning when the grant is formally awarded and ends after two years to ensure that the federal funds are spent promptly. However, a grant may receive a maximum extension to five years. Allowable costs shall be determined in accordance with the 7 CFR 3016, "Uniform Administrative Requirements for Grants and Cooperative Agreements to State and Local Governments," and any amendments to this regulation.

Donations of land or interests in land must be documented to count as part of the nonfederal cost-share. The title does not need to be transferred to the State or federal government in order for the donation to qualify as cost share. However, if in the future, the donated lands are conveyed or the rights or title are modified in a way that is inconsistent with the purposes of the FLP then the State must restore the cost share value dedicated in the grant agreement. The value of donations may be included as part of the nonfederal cost-share if all of the following are met:

1. The donation contributes to the objectives and priorities of the State FLP as set forth in the AON;
2. All or part of the tract being donated must be within the boundaries of an FLA, and may include National Park,

National Forest, National Wildlife Refuge, or other Federal land boundary, or within the boundaries of an area designated through an analogous State program with goals compatible with the FLP and be within an FLA;

3. The donor documents their desire that value of the interests may be used as cost share for the FLP project;
4. The donation of land or an interest in land must contain perpetual covenants to assure that the tract will be managed in a manner compatible with the goals for which the FLA was established;
5. The donee (holder of donated rights) is a unit of government or a non-profit conservation organization (land trust) that meets the eligibility requirements for holding a conservation easement established by the Internal Revenue Service and has as its purpose the management of lands or interests in land consistent with FLP purposes;
6. If the donation is in the form of a conservation easement then the deed needs to contain a provision that directs all of the easement holder's proceeds from a subsequent sale or exchange of interests in land be used in a manner consistent with the conservation purposes identified for the subject interests in lands;
7. The respective portion of the donation must not have been previously credited towards any Federal program's nonfederal cost share; and
8. The State lead agency approves the donation as contributing to the cost-share.

Acquisition of Lands or Interests in Lands

FLP acquisitions may be outright full fee purchases, or acquisition of development rights or other rights conveyed through a conservation easement. Except in the case of a full and complete donation of land or an interest in land, if any Federal funds are used in the acquisition of Forest Legacy tract the following shall apply:

1. Federal appraisal standards must be met, including appraisal review by a qualified Review Appraiser;

2. The landowner must be informed in writing of the market value and that sale of the property is strictly voluntary;
3. The landowner must be notified in writing that the property will NOT be purchased if negotiations do not result in amicable agreement;
4. Federal payment to the landowner for lands or interests in lands is not more than the market value as determined by an appraisal meeting Federal appraisal standards;
5. The title acquired must be free of encumbrances inconsistent with the purposes of the FLP. Title insurance may be secured for the full value of the encumbered property, but is not an alternative to an acceptable title; and
6. If relocation is involved the requirements in the Uniform Relocation Assistance and Real Estate Property Acquisition Policies Act of 1970 (PL 91-646 or 42 U.S.C. 4601) must be followed. The FS will be advised in advance of any acquisition involving relocation.
7. In the case of acquisition of interested in lands, the development of a Forest Stewardship Plan or multi-resource management plan that has been approved by the landowner and the State Forester or designee and Baseline Documentation Report shall be prepared prior to project closing (See Appendix J for sample content and references).

All FLP acquisitions of lands or interest in lands are perpetual and therefore run with the land. Although any remaining interests held by the landowner may be subsequently conveyed, future owners are still bound by the terms and conditions of the conservation easement. At the same time, future owners shall retain full control of the rights that are not acquired by the FLP, and shall be subject only to those restrictions that the present landowner has conveyed to the Federal, State, or local government. Compatible nonforest land uses (cultivated farmland, pasture, grassland, shrubland, open water, and wetlands) are desirable land uses in many FLAs. FLP funds should not be used for any property not meeting the State's definition of forested land in the AON, unless there is a written plan scheduling reforestation or afforestation. Programs to conserve farms, ranches and similar land uses may be used in conjunction with the FLP to protect properties where there are mixed forest and compatible nonforest uses.

Conservation easements are required to contain language pertinent to the purpose of the FLP and a reversionary provision to ensure the conservation investment of FLP into the future (Example clause language are found in Appendix I). During the development of tract specific conservation easements, a determination will be made as to whether the acquisition of mineral rights, prohibition on reserved areas, or an exclusion of the area that does not comply with FLP, would be necessary in order to protect the other rights that are being considered for acquisition. In some situations, it may be impossible to protect environmentally important forest areas pursuant to the purpose of the FLP without acquiring the mineral rights.

The FLP conservation easement holder (Federal, State or local government) is responsible to assure that baseline documentation contains all the information necessary to monitor, manage and enforce the easement. Where the conservation easement is a tax-deductible gift, and the owner retains rights to the property, the Internal Revenue Service (IRS) holds the donor responsible for providing sufficient baseline data "to establish the condition of the property at the time of the gift." (See Treas. Reg.§1.170A-14(g)(5)(i)). However, this does not eliminate the FLP need for baseline documentation.

Baseline documentation describes or depicts a tract of land and its attributes on the day it becomes restricted by an easement. This documentation is required on all FLP tracts and is completed prior to project closing. Documentation of the property should include a map of the area drawn to scale showing all existing man-made improvements or incursions such as roads, buildings, fences or gravel pits; an aerial photograph of the property taken as close to the date the property is restricted as possible; and on-site photographs, especially of significant features. The above should be accompanied by narrative descriptions of tract attributes and other pertinent information.

States and landowners are encouraged to display the official FLP signs on the FLP property using the signs in accordance with Appendix K. The posting of FLP tracts helps promote public awareness, recognition and support for the program. Landowner permission should be secured before posting any signs. Costs associated with sign posting can be covered by FLP project or administration grants or States may use such expenses as FLP cost share. Signs should be inspected during the annual monitoring of the FLP tract and repaired when in poor condition. FLP sign art and program logos may be used by FLP partners for items that contribute to the purpose of awareness (e.g. brochures, workshops, outreach efforts, posters, FLP information packets, web sites etc.)

Appraisal and Appraisal Review

The FLP policy on appraisal is that all FLP acquisition of land or interests in land using Federal funds must comply with Federal appraisal standards contained in the publication entitled "Uniform Appraisal Standards for Federal Land Acquisitions: Interagency Land Acquisition Conference, 2000," as amended or updated. Appraisals and appraisal reviews may be conducted by any qualified appraiser meeting the minimum standards outlined in Appendix H.

The FLP will ensure high quality appraisal service and accountability to the program by:

- o Annual planning and coordination of appraisal work to allow for efficient allocation of resources.
- o Requiring checks and balances:
- o States will ensure that qualified appraisers trained and competent in appraisal, appraisal review and knowledgeable of Federal standards will be used. The State may use State, contract or Federal appraisal or review services to meet this requirement.
- o States or the FS will review contract appraiser qualifications as stated in Appendix H before they are employed to conduct a FLP project appraisal or review.
- o The appraiser and identified review appraiser will engage in an initial consultation before the project appraisal takes

place. The review appraiser will develop project specific appraisal instructions for the appraiser as a result of this consultation.

- o The FS will conduct spot checks of appraisal reviews to ensure quality and accuracy.
- o Forest Legacy funds can only be used to purchase lands and interests in land after the appraisal review confirms that the appraisal meets the Uniform Appraisal Standards for Federal Land Acquisitions. It is recommended that an offer not be made until the appraisal review is approved.

Conservation Easement Monitoring, Management, Record-Keeping & Enforcement

The governmental entity holding title to interests in land acquired under the FLP shall monitor and manage those interests in perpetuity. The holder may delegate or assign monitoring, management, and enforcement responsibilities over lands and interests in lands acquired under the FLP only to other Federal agencies or State or local government entities.

Such delegation or assignment of responsibility shall be documented by a written agreement. The governmental entity responsible for monitoring, management and enforcement of the conservation easement may in turn delegate or assign management and monitoring authority to other parties, to include land trusts, conservation groups, and other governmental entities. Such delegation or assignment of authority shall be adequately documented and the FS shall be notified. The FS shall approve agreements involving any interests in lands held by the Federal Government prior to such delegation or assignment. Once interests in lands are acquired, the State lead agency, FS, and others as appropriate, may negotiate tract-specific Memorandums of Understanding (MOU) as necessary to specify management and monitoring responsibilities for the interests in lands.

Optimal management and monitoring of tracts in FLAs is based upon partnerships between landowners, private non-profit organizations owning or managing lands, and State and Federal officials. Land trusts and other private organizations will continue to manage and monitor their own easements and lands within designated FLAs, and while they may not manage government-owned interests in lands under the FLP, they may cooperate with or contract for monitoring and implement specific management activities. Management of federally owned interests in lands is reserved to the FS, but may be assigned to State or local governments, or another Federal agency through mutual agreement. Although delegable, enforcement actions for easements will generally be conducted by the easement holder, i.e., the State or the Federal Government.

Monitoring FLP conservation easements shall occur periodically, but not less than annually. Monitoring consists of visual inspection of the property, documented by a written report to explain the condition of the property at time of inspection. Any material departure from the baseline documentation report or Forest Stewardship Plan should be noted.

The easement holder should immediately address any violation of the conservation easement with the landowner. The landowner should have the opportunity to correct the breach. After a reasonable time period (e.g. 30 days), if the breach is not corrected, enforcement action may be taken, including but not limited to, legal means.

The unit of government holding the conservation easement has the initial responsibility to enforce the conservation easement. The State or easement holder shall promptly notify any future FLP tract owner of the FLP and the origin and requirements of the conservation easement. The Forest Stewardship Plans covering the tract shall be reviewed periodically and updated as needed. If there is a change in land ownership, then the Forest Stewardship Plan needs to be reviewed, and updated as needed.

Landowner Participation

Landowner participation in the program is voluntary and consists of two elements:

1. Conveyance of lands and interests in lands to achieve the purpose of the FLP;

2. Preparation and periodic updates of a Forest Stewardship Plan or a multi-resource management plan. The landowner and the State Forester or designee must approve the plan prior to signing the acquisition of the easement. The plan shall include provisions to meet land conservation objectives of the FLP.

   The plan shall be kept current and updated as needed. Modifications of the plan must be agreed to by the State lead agency. A plan is not needed if the lands are purchased in fee. (See Appendix F for sample content of a Forest Stewardship Plan) Landowners may submit an application and property information (See Appendix E) to the State lead agency to enroll their land or interests in lands in the FLP according to the process described in the AON. All owners of eligible forestlands within the designated FLA, and meeting the minimum Eligibility Criteria or other application requirements described in the AON, are eligible to submit an application.

For a landowner to participate in the program, it is not required that their tracts be completely forested. (see definition of "Nonforest Uses" and "Reserved Areas") However, priority will generally be given to tracts that are currently forested or are identified to be forested in the landowner Forest Stewardship Plan or multi-resource management plan.

The FLP respects the rights of private property holders. Under no circumstances shall the right of eminent domain be used for the unwilling "taking" of any private property rights. Traditional forest uses such as forest management activities, including timber management, and outdoor recreation opportunities are deemed consistent with purposes of the FLP and are encouraged on FLP tracts when consistent with the State's AON and the conservation purposes for FLP tract acquisition.

The FLP adheres to language contained in Section 14 of the CFAA, Statement of Limitation: "This Act shall not authorize the Federal Government to regulate the use of private land or to deprive owners of land of their rights to property or to income from the sale of property, unless such property rights are voluntarily conveyed or limited by contract or other agreement.

This Act does not diminish in any way the rights and responsibilities of the States and political subdivisions of States." Purchase or donation of rights does not relieve landowners of regulations that would otherwise apply.

The FS has no jurisdiction to make tax determinations or render advice as to the tax implications of transactions. Since tax implications differ from person to person, landowners should be encouraged to seek independent counsel from local assessors, tax lawyers, or accountants.

Land Trust Participation

Land trusts are nonprofit organizations that protect land by working with landowners wishing to donate or sell fee title or conservation easements to maintain conservation values associated with the land. Land trusts can have an important role to play in the FLP. The following considerations apply to land transactions between the Federal Government/States and land trusts:

1. Land trusts cannot execute contracts for acquisition of interests in lands on behalf of the Federal /State Government. Land trusts may work as intermediaries for eventual Federal/ State acquisition, but without an accepted land purchase option and contract with the FS there is no guarantee of Federal acquisition. No pass-through transactions shall be done without prior consultation with the FS/State.
2. With approval of the State lead agency, the FS, the land trust or the donating landowner, lands and interests in lands acquired by land trusts (pursuant to Final Guidelines Part 1, Section XIII) may be counted toward the nonfederal cost-share contribution, provided that the interests in lands permanently contribute to the FLP.
3. If a land trust proposes a pass-through transaction to the FLP it must assure that terms and conditions in the deed or conservation easement are reviewed and approved in advance by the State lead agency and/or the FS.

4. The monitoring of easements within FLAs may be performed by land trusts in accordance with the umbrella MOU for the FLP in that State and individual MOUs for specific tracts established between the State and the land trust organization.
5. Other appropriate and beneficial roles of land trusts in relation to the FLP may include: participation on the SFSCC; recruitment and facilitation of FLP projects; buyer of tracts or easements of proposed, but unfunded FLP projects; facilitators of local FLP efforts; and performing tract monitoring and management activities.

PART 2-STATE GRANT PROGRAM

The State lead agency elects the State grant option of the FLP, in writing, to the appropriate FS Region/Area/IITF.

When a State elects the State grant option, all FLP acquisitions shall be transacted by the State with title vested in the State or a unit of State or local government. There are two exceptions:

1. Donations where the donor may wish to make a donation to a land trust, local, or Federal Government and the donee agrees to accept the donation, and to manage the lands or interests in lands in perpetuity for FLP purposes; and
2. At the request of the State and at the discretion of the FS, the FS may acquire individual tracts or multiple tracts within a specified FLA, with title vested in the U.S. in accordance with Part 3 of these guidelines.

Grants

If a State elects the optional State grant option, the FS will provide a Federal grant to the State to carry out the FLP, including the acquisition by the State of lands and interests in lands. Grants must be consistent with the uniform administrative requirements established in 7 CFR 3016.

States will generally be reimbursed for costs incurred with cash advances limited to the minimum amounts needed and timed to be in accord only with the actual, immediate cash requirements of the State in carrying out the FLP. The timing and amount of cash advances shall be as close as is administratively feasible to the actual cash outlay by the State for direct program costs and the proportionate share of any allowable indirect costs.

Conditions of the Grant

1. States must submit annual performance and financial status reports. A final performance report and financial status report are required prior to close out of the grant.
2. Funds appropriated for the FLP shall not be included in consolidated-payment grants made under authority of Section 12 of the CFAA.
3. The State shall maintain current and complete financial records in accordance with requirements contained in the latest Federal Aid Manual and OMB Circular.

Eligible Activities

The following activities are eligible uses of funds granted to States for the FLP; however, in most cases costs incurred prior to issuance of the grant cannot be reimbursed:

1. Purchase of lands or interests in lands from willing sellers for inclusion in the FLP;
2. Facilitation of donations of lands or interests in lands to a qualified and willing donee for FLP purposes;
3. Program administration expenses limited to indirect costs and direct acquisition related expenses for lands and interests in lands acquired under Forest Legacy authority;
4. Establishment and documentation of baseline conditions and development of a Forest Stewardship Plan for a conservation easement; and
5. AON Planning and amendment.

The following uses of Forest Legacy funds are not allowed as part of a State grant:

1. Management of acquired lands or interests in lands including, monitoring of conservation easements,
2. Enforcement actions, and

3. Payment for appraisals of donated property when the donation represents the full and total value.

Availability of Funds

Project funds for any fiscal year shall be available to the State for two years from the time they are obligated in a FS grant to the State in order to insure that Federal funds are spent promptly to acquire FLP projects.

However, a grant may have a maximum duration of five (5) years to allow for nonfederal cost sharing to occur. During the 5-year life of the grant, it can be amended annually, as needed, and funds from a new fiscal year added to the grant, consistent with the requirement that the funds be expended within two years of the time of obligation. In no case can funds be obligated or expended beyond the 5-year life of the grant.

Acquisition of Lands and Interests in Lands by States

All Forest Legacy acquisitions including the acquisition of lands or interests in land shall be made in accordance with Federal appraisal and acquisition standards and procedures. The interests in land acquired for Forest Legacy shall be adequate for Forest Legacy purposes and be perpetual. Title to such lands or interests in lands will be vested in the State or unit of State government. These lands or interests in lands will be managed and administered for goals consistent with Forest Legacy conservation purposes by State agencies or their assigns. The State agencies are responsible for all monitoring and management of conservation easements and management of fee simple properties.

Lands and interests in land located within a FLA and simultaneously within other Federal boundaries (e.g. national forest, national park, or national wildlife refuge) are eligible for the FLP provided that the responsible Federal agency concurs with the FLP State acquisition. If a State has passed legislation that extinguishes claims to or restrictions on real property, the State shall use all available authorities, including that of acting as an agent of the U.S., to achieve the purposes of section 7(K)(2) of the CFAA.

Reversion of Funds for Forest Legacy Inconsistency

In the event it is determined, by the State lead agency, that it is no longer desirable to hold lands or interests in lands acquired with Federal funding and those lands are conveyed, exchanged, or otherwise disposed of, after providing notice to the FS, the State shall:

1. Reimburse the FS for the current market value in proportion to the original Federal investment; (said reimbursements to be used to further the purposes of the FLP); or
2. Exchange for other FLP eligible lands or interests in lands of at least equal market value and of reasonably equivalent location, with public purposes that equal or exceed those of the disposed tract, with FS approval.

Items 1 and 2 identified above must be included in deeds or conservation easements of all FLP tracts as well as in the FS grant to the State. Appendix I includes suggested language for conservation easements and deeds.

PART 3-FEDERAL ACQUISITION PROGRAM GUIDELINES

Federal Acquisition Process

In the furtherance of the purposes of the FLP, the State lead agency with involvement of the SFSCC and the FS will review property owner applications, prioritize tracts, obtain State approval, and submit properties to the appropriate FS Region/Area/IITF for funding. Upon approval for funding, the FS will proceed to acquire from willing sellers conservation easements and/or other interests in land including fee acquisition. Federal Acquisition Procedures must be followed when Federal funds are used to complete an acquisition of land or interests in land using FLP authority. They are:

1. Federal appraisal standards must be met;
2. The landowner must be informed of the market value and that sale of the property is strictly voluntary;
3. The landowner must be notified in writing that the property will NOT be purchased if negotiations do not result in an amicable agreement;

4. Federal payment to the landowner for lands or interests in lands is not more than the market value determined under #1;
5. Assure title is free and unencumbered relative to the purposes of the FLP; and
6. If relocation is involved the requirements in PL 91-646 (42 U.S.C. 4601) must be followed and the FS must advise the landowner prior to the acquisition.

Certain lands are not eligible for the Federal ownership option under FLP authority because other authorities and funding sources are available for acquisition of lands or interests in lands within these federally established areas. These include lands or interests in lands located within National Forests, National Parks, National Wildlife Refuges, or other Federal Government boundaries. Proximity to Federal lands or the inclusion of Federal lands within a proposed FLA does not disqualify an area for program eligibility.

Federal laws governing public lands do not apply to private property rights not acquired by the Federal Government from willing private landowners. Interests in lands retained by private landowners, not conveyed to the Federal Government under the FLP, are subject to the same requirements of the Endangered Species Act (ESA) that existed prior to their participation in the FLP. Conveyance of interests in lands to the Federal Government neither enhances nor diminishes the landowner's responsibility under the ESA. Any interests in lands acquired by the Federal Government under the FLP shall be subject to the same requirements of the ESA as are other Federal lands.

Memorandum of Understanding (MOU) for Coordination of the FLP

An MOU will be used to coordinate the FLP where Federal acquisition option resulting in Federal ownership of FLP acquisitions occurs. The MOU will define and facilitate partnerships between the State lead agency, FS, and other participating entities in implementing the program, acquiring interests in lands, and sharing the costs of the program. The MOU shall determine how costs are shared between parties, including administrative, management, monitoring, and capital improvement expenses. The terms of a MOU will determine which party is responsible for costs incurred following the project's five-year cost-share write off period. If individual Forest Legacy tract-MOUs are needed, they become an addendum to the State level "umbrella" MOU. The umbrella MOU between the State lead agency and the FS shall be developed following the Secretary's approval of the State's AON and the establishment of the State's FLP.

The FS/State MOU is for the purpose of specifying roles and responsibilities for implementing the program, and may address the following items:

1. Costs and Funding:
    a. Identify direct and indirect costs expected to be incurred in establishing the FLP, and acquiring and administering interests in lands during the first five years of the program. Revise or renew these cost estimates as appropriate.
    b. Identify and propose sources of cost-share matches.
2. Planning:
    a. Document the amount of work required to complete the AON and identification of FLAs.
    b. Define a process for revising existing landowner Forest Stewardship Plans, or multi-resource forest management plans.
    c. Identify how specific tract acquisition needs and priorities shall be established by the State.
3. Acquisition:
    a. Identify who is responsible for title work, appraisals, surveys, and similar pre-acquisition work.
    b. Define a process for determining the value of donated interests in lands.
4. Management:
    a. Define responsibilities for management of interests in lands acquired or dedicated to the program.

   b. Identify possible activities needed to enhance, restore, or maintain resources to meet the intent of the program and general responsibilities in carrying out such activities.

5. Administration:
   a. Estimate the staff-work required to implement the Program.
   b. Define responsibilities for processing applications to the FLP.
   c. Establish procedures for monitoring and enforcement the terms of reserved interest deeds and easements and identify who will be responsible.
   d. Identify responsibilities for periodic reports summarizing the achievement of FLP goals in the State.

Payment in Lieu of Taxes (PILT)

Where the Federal Government under the FLP acquires lands in fee, the Federal Government will pay PILT to the local taxing authority. No PILT will be paid on conservation easements.

Transition to State Grant Option Program

If a State elects the State Grant Option, and there are active cases being pursued by the FS, all parties (FS, State, and landowner) may agree to transfer the case to the State. If agreement to transfer is reached, then the value of the lands or interests in lands comprising the project may be transferred to the State by a FS grant. To facilitate projects transferred to the State, the FS may provide the State with copies of any appraisals, appraisal reviews, title reports, option contracts and other pre-acquisition materials for lands that have been under negotiation by the FS.

FS completes FY 2005 Recommended Project Briefing Book State Lead Agencies receive applications for FY 2005 FLP projects (some States have an open application process) State Forest Stewardship Coordinating Committees meet to evaluate and approve applications for FY 2005 funding according to the criteria identified in the State AON.

FS regional units consult with States to create a FY 2005 "Project Opportunity List" that includes all projects that they are requesting funding for in priority order.

FS regional units develop a FY 2005 "Project Recommendation List" based on regional criteria, State project lists and project briefings and submit to FS Washington Office.

FS regional units submit FY 2005 "Project Opportunity List" to FS Washington Office.

FS conducts National Review Panel representing FS and State regions to prepare a prioritized FY 2005 "Forest Legacy Project.

Recommendation List" for submission to Administration.

FS submits FY 2005 FLP Project Recommendation List to the Administration.

FS notifies the House and Senate Appropriations Committee on the project list after it clears the Administration.

Congressional committees discuss and pass annual Appropriations for FY2005 projects.

EXAMPLES OF COST SHARE CALCULATIONS

Equation for Calculating Cost Share Requirement (Federal FLP Share) X (0.333) = the minimum Non-Federal Contribution

OR

(Total Project Costs) X (0.75) = the maximum Federal Contribution

Principals to Guide Calculating the Cost-Share Requirements

- o To calculate the cost share requirement, the Program Manager should use the Federal FLP contribution, and not the total project costs.
- o The cost share requirement should be at least 33.3% of the total Federal FLP contribution towards the project, which will equal at least 25% of the total FLP project (Federal FLP contribution plus cost share).
- o The Federal contribution (Forest Service's FLP plus all other Federal contributions) cannot exceed 75% of the total project

costs (all cost requirements to complete the project, including Federal and non-Federal contributions).

- o The non-Federal cost share portion cannot be used as cost share for another Federal program that also requires a cost share.

Example 1-The FLP is going protect Jane Smith's 3,000 acres tract. The total cost of protecting that land is $1 million.

Total Project Costs	Federal FLP Contribution	Non-Federal FLP Contribution	Other Federal Contribution	Other non-Federal Contribution
$1,000,000	$750,000	$250,000	$0	$0

- o Federal contribution cannot exceed $750,000; therefore, the Federal contribution is not greater than 75% of the total project costs.
- o The non-Federal cost share requirement is at least $250,000; therefore, FLP funds are adequately cost shared.

Example 2-John Doe Ranch is planning to conserve 6,500 acres of land. The total cost of protecting the land is $4 million. The Federal contribution, through FLP, will be $1,000,000, and the non-Federal contributors will provide $3,000,000, which includes a cost-share component for the FLP.

Total Project Costs	Federal FLP Contribution	Non-Federal FLP Contribution	Other Federal Contribution	Other non-Federal Contribution
$4,000,000	$1,000,000	$333,000	$0	$2,667,000

- o Federal contribution cannot exceed $3,000,000; therefore, the Federal contribution is not greater than 75% of the total project costs.
- o The non-Federal cost share requirement is at least $333,000; therefore, FLP funds are adequately cost shared.

Example 3-ABC Tree Company is planning to conserve 8,300 acres of land. Both theForest Service's FLP and the U.S. Fish and Wildlife Service (FWS) are contributing funds toward the project. Non-Federal money has been secured to cover the non-Federal cost share requirements for the FLP and FWS requirements, as well as to pay for additional project costs.

Total Project Costs	Federal FLP Contribution	Non-Federal FLP Contribution	Other Federal Contribution	Other non-Federal Contribution
$4,000,000	$1,000,000	$333,000	$1,000,000	$1,667,000

- o Federal contribution cannot exceed$3,000,000; therefore, the Federal contribution is not greater than 75% of the total project costs.
- o The non-Federal cost share requirement is at least $333,000; therefore, FLP funds are adequately cost shared.
- o FLP cost share component cannot be the same as the FWS cost share component.

INFORMATION TO FACILITATE LANDOWNER PARTICIPATION

Landowners who wish to participate in the program may be asked to provide the following information.

1. Name, address and phone number of applicant landowner.
2. All other owners of record for this tract, and their addresses.
3. Name, address and phone number of authorized agent representing landowner(s) if applicable.
4. Location of property.
5. If the landowner intends to reserve rights to forestry uses or other resource management activities, a copy or reference to the State approved landowner Forest Stewardship Plan or multi-resource management plan.
6. List of the significant scenic, natural, recreational, wildlife, timber and other resource values contained on the property.
7. Identification of all dams, dumps or waste disposal sites on the property.
8. Signed statement giving the FS and State lead agency permission to enter the property for review and appraisal purposes.

9. Legal description.
10. List any encumbrances or liens existing on the property including, but not limited to contracts, leases, or outstanding rights not of record.
11. Copy of plat or survey map of the property, if existing. If only a portion of the property is being offered, identify it on a plat showing the portion offered in the context of the entire tract.
12. Tract acreage and total number of acres of forests and cleared/open land.
13. List of existing permanent improvements on the tract, including houses, barns, lakes, ponds, dams, wells, roads, and other structures, and total number of acres occupied by improvements.

SAMPLE CONTENT OF A FOREST STEWARDSHIP PLAN

Below is information from the Forest Stewardship Program's National Standards and Guidelines. Please also refer to the Forest Stewardship Program's Planning for Forest Stewardship: A Desk Guide as well as States' Statewide Forest Stewardship Plans for additional information on Forest Stewardship Plans.

Landowner Forest Stewardship Plans must:

- o be prepared or verified, as meeting the minimum standards of a forest stewardship plan, by a professional resource manager.
- o identify and describe actions to protect, manage, maintain and enhance relevant resources listed in the law (soil, water, range, aesthetic quality, recreation, timber, water, and fish and wildlife) in a manner compatible with landowner objectives.
- o be approved by the State Forester or a representative of the State Forester.
- o involve the landowner in the plan development by setting clear objectives and should understand clearly the completed plan.

A well prepared plan will:

- o Clearly state landowner objectives.
- o Have a cover page.
- o Provide for authorship and/or signature lines within the document.

The plan preparer should consider and evaluate resource elements present and include a brief description of those that are applicable and their importance to the ownership. Resource elements to be considered are:

- o Soil Interpretations
- o Water
- o Range
- o Aesthetic Quality
- o Recreation
- o Timber
- o Fish
- o Wildlife
- o Forest Health
- o Archeological, Cultural and Historical Sites
- o Wetlands
- o Threatened and Endangered Species

Management recommendations, or where appropriate, alternative strategies should be provided for those resource elements described. Prescriptions or treatments should be integrated and stand or site specific. An ownership map drawn to scale, or photo, to include vegetation cover types, stream and pond location with a legend will enable the landowner to implement the plan. Landowners' understanding may be improved by including activity summaries and appendices.

Appendices might include:

- o Description of assistance available and incentive programs
- o Educational materials

- o A glossary of terms
- o An explanation of applicable Federal, State and /or county regulatory programs, especially as they apply to:
- o Archeological, cultural and historical sites.
- o Wetlands.
- o Threatened and Endangered Species.

These last three items are covered by legislation other than the Cooperative Forestry Assistance Act of 1978, as amended by title XII of the Food, Agriculture, Conservation and Trade Act of 1990 (16 U.S.C. 2101, et seq.), but must be considered for Federally funded programs. The professional resource manager should discuss the Forest Stewardship Plan with the landowner, following completion, to assure understanding.

REAL ESTATE RECORD KEEPING

Since Forest Legacy acquisitions are perpetual, record keeping is important. Each State shall maintain permanent records for all Forest Legacy properties. The following information is recommended to be maintained by the conservation easement holder:

A. Landowner information (name, address, phone)

B. Nomination form (including notification to landowner that property will not be purchased if negotiations do not result in amicable agreement)

C. Landowner Inspection Consent Agreement

D. Baseline documentation

E. Option agreement

F. Deed of Conservation Agreement

G. Additional warranty deeds, covenants, restrictions

H. Title Insurance Policy

I. Appraisal

J. Appraisal review

K. Forest Stewardship Plan or equivalent

L. Notification of county or local government

M. Closing statement

N. Copies of check or documentation of EFT or other form of payment

O. Copies of grant reimbursement or expenditure

The following items should also be maintained as part of the record:

1. Landowner correspondence
2. Evaluation criteria
3. Tracking/documentation of negotiation steps
4. State Forest Stewardship Coordinating Committee recommendation
5. Press release
6. Monitoring records/history

REQUIRED QUALIFICATIONS OF AN APPRAISER OR REVIEW APPRAISER

A. Appraiser-In order to be a qualified appraiser for purposes of FLP appraisals, an individual must be:

1. a Federal land acquisition agency staff appraiser who
    a. is certified as a general appraiser in compliance with OMB Bulletin 92-06, and
    b. has completed training in application of the December 2000 edition of Uniform Appraisal Standards for Federal Land Acquisitions(UASFLA)* approved for appraiser continuing education credit in the State where the appraiser is certified, or
2. a nonfederal staff or fee appraiser who
    a. is certified as a general appraiser in the state where the appraised property is located, or can obtain reciprocity or a temporary practice permit in the state where the appraised property is located, and
    b. has, within the past 10 years, completed at least the minimum classroom hours of non-duplicative education prescribed for the certified general real

property appraiser classification by the Appraisal Standards Board of The Appraisal Foundation, and

c. has completed at least 12 self-contained or summary appraisal reports of properties similar in scope and complexity to the appraised property in the preceding three years, and

d. has completed training in application of the December 2000 edition of Uniform Appraisal Standards for Federal Land Acquisitions approved for appraiser continuing education credit in the state where the appraiser is certified.

The qualified appraiser shall prepare an appraisal report in compliance with the Uniform Appraisal Standards for Federal Land Acquisitions and supplemental written appraisal instructions issued by the client. Federal land acquisition agencies are the member agencies of the Interagency Land Acquisition Conference.

B. Review Appraiser-In order to be a qualified review appraiser for purposes of FLP appraisals, an individual must be:

1. a Federal land acquisition agency staff appraiser who
   a. is certified as a general appraiser in compliance with OMB Bulletin 92-06, and
   b. holds specific delegated authority to review and approve or recommend appraisals for agency use, and
   c. has completed training in application of the December 2000 edition of UASFLA* approved for appraiser continuing education credit in the State where the reviewer is certified, or
2. a nonfederal staff or fee appraiser who:
   a. is certified as a general appraiser in the State where the appraised property is located, or can obtain reciprocity or a temporary practice permit in the state where the appraised property is located, and

b. has, within the past 10 years, completed at least the minimum classroom hours of non-duplicative education prescribed for the certified general real property appraiser classification by the Appraisal Standards Board of The Appraisal Foundation and at least 32 classroom hours of approved training in appraisal review, or otherwise demonstrates competency in appraisal review in compliance with the Competency Rule of the Uniform Standards of Professional Appraisal Practice (USPAP), and

c. has completed at least 12 self-contained or summary appraisal reports of properties similar to the appraised property in the preceding three years or at least 12 technical appraisal review reports for appraisal reports of properties similar in scope and complexity to the appraised property in the preceding three years, and

d. has completed training in application of the December 2000 edition of Uniform Appraisal Standards for Federal Land Acquisitions approved for appraiser continuing education credit in the state where the reviewer is certified.

The qualified review appraiser shall prepare a technical appraisal review report that includes a determination of whether the appraisal report under review complies with the Uniform Appraisal Standards for Federal Land Acquisitions. Federal land acquisition agencies are the member agencies of the Interagency Land Acquisition Conference. *The seminar, Federal Land Exchanges and Acquisitions: Appraisal Issues and Applications, offered by the American Society of Farm Managers and Rural Appraisers and the Appraisal Institute is the only acceptable substitute for UASFLA training.

THE HISTORY GLOBAL ENVIRONMENT

Controversy about forest destruction has been the Bank's Achilles Heel for nearly two decades. The flagship disasters in Brazil and Indonesia in the 1980s focused on the role of the Bank as forest destroyer. The subsequent attempts by the Bank to do good forestry through the TFAP (Tropical Forest Action Plan) proved a further public relations disaster for the Bank. The TFAP was an ambitious plan launched by the World Bank and other international actors in the mid-1980s to mobilize US $ 8 billion over a period of five years. While the official goal was to curb deforestation, the TFAP's planned investments were very much geared towards industrial logging activities. It was the uniting of ecological justice NGOs in opposition to the TFAP that created the WRM. The TFAP debacle, which the WRM triggered, and broad-based criticism of environmentally destructive forestry loans for West Africa forced a rethink of forest policy in the Bank in the late 1980s and early 1990s. The WRM network was influential in this policy review, which resulted in a new World Bank Forest Policy in 1991.

The 1991 Policy was an important step forward for the Bank. It committed the Bank to a new approach to forests including commitments to:

- assess the likely impacts on forests of macro-economic and non-forest sector loans: SAPs, agricultural projects, dams, mines etc.
- take into account the interests of communities in establishing plantations
- involve forest-dwelling people in establishing protected areas
- avoid funding logging in primary moist tropical forests
- condition other lending for forest operations on the inclusion of measures to safeguard the interests of forest-dwellers
- experiment with new projects with forest-dwellers including in promoting non-wood products.

In 1994 the Bank carried out a Forest Policy Implementation Review, partly as a result of insistence of NGOs and a supportive US Executive Director. This was a rather unsuccessful stock-taking exercise, although the scope given for NGO participation in the process was a step forward in Bank procedures. A half-hearted attempt by some Bank staff to have the proscription on funding of logging in primary moist tropical forests was lifted by sustained

NGO pressure. The WRM was especially critical of the failure of the review to look at:

- the cross-sectoral impacts of Bank lending on forests
- the lack of staff incentives to observe the policy, especially with respect to adequate participation
- the social impacts of Bank lending related to forests, especially in terms of impacts on women, indigenous peoples and land tenure, or measures to address these impacts (see annex 2).

The Bank admitted that these were real deficiencies in its review but that it was obliged to report to the World Bank's Board of Executive Directors (EDs) that year and so had no time to fill the gaps. The WRM heavily lobbied the EDs at the Bank not to accept the Review as it stood. In the end they did accept it on condition that a follow up review was carried out in 1997. After many delays this revised review is now underway.

What's New?

Since 1991, there have been a number of new developments which make this a significant moment for the Bank and for forests:

- The Bank is now active in transition countries and is starting forestry projects there: the 1991 policy was essentially focused on tropical forests only. This might result in greater pressure on forests and processes similar to those occurring in the South.
- Bank lending is increasingly through programme lending and is less project linked.
- The Bank seeks to direct more funding through the International Finance Corporation (IFC), its branch which lend directly to the private sector. Other support of private sector investment is channeled through the World Bank's MIGA (Multilateral Investment Guarantee Agency, which provides political risk insurance to private businesses.
- The Bank is seeking to become the financial channel for forestry funding linked to the Kyoto Protocol (Clean Development Mechanism), which might result in massive

substitution of natural forests by monoculture tree plantations.

- Global negotiations on forests are nearing consensus on some issues which we believe to be less relevant and which might form the basis for a forests convention. The Bank is seeking a role in implementation of any international agreement which arises.The issues on the table at the global negotiations are questions related to 'national sovereignty', 'sustainable forest management', 'criteria and indicators for sustainable forest management', 'National Forest Programmes', 'Promotion of Trade in forest products', 'voluntary self-regulation' etc. In contrast, the issues we want to be addressed are indigenous peoples rights, participatory agrarian reform, community control of forests, reforms in aid, regulation of transnational investments, limits on cross sectoral forces pressing on forests, halt in illegally produced forest products, the equivalence of international laws on human rights and the environment with international laws on trade.
- A new, feisty set of World Bank forestry employees are much more aware of the role of forestry in national political economies but remain convinced that, done right, forestry can be socially and environmentally sustainable and good for national development. They are seeking to use the heavy conditionality of structural and sectoral adjustment lending to reform developing country forestry policies and practices. Critical facts are simply being ignored such as the lack of evidence that large-scale timber extraction can be sustainable, meet local people's needs or contribute meaningfully to poverty alleviation, while its political effects in concentrating power in the hands of often corrupt elites is a major obstacle to national development.

SECOND GUESSING THE BANK'S VISION

There is still a great deal of debate within the Bank about the appropriate way forward when it comes to forestry. A vision can be seen to be emerging, however.

This vision seems to include the following elements:

- Tropical forests will be all but gone in 50 years if current trends continue
- Staff in developing country forestry departments are becoming less not more adequately trained.
- Key reforms are needed to promote collaborative forest management and reformed forest governance.
- Curbing forest piracy is essential.
- Sustainable Forest Management (at least sustaining flows of some key goods and services) is possible if there is a level playing field in terms of standards, regulations and costs.
- The Bank should therefore support a Global Forests Convention.
- Rates of return on SFM are however not competitive with the return on capital expected by private investors today, especially if the costs of protection and provision of ecological services are included.
- State regulation of forestry and laws to maintain PFE (Permanent Forest Estate) are thus crucial.
- Collaborative forestry options are very useful but not a panacea.
- Logging is not really responsible for forest loss.
- Certification should be promoted.
- The prohibition on Bank funding of logging in primary moist tropical forests should be lifted.
- The Bank should integrate forestry issues into Country Assistance Strategies and into structural adjustment conditionalities.

The FPIRS Process

The process of the current FPIRS review has already been substantially changed owing to NGO lobbying. An initial proposal circulated at the very end of 1997 was to carry out a hurried implementation review and strategic planning process in the early months of 1998. This was dropped when senior Bank management agreed with NGOs that it was hasty and ill-thought out.

As conceived now an 'independent' FPIR will be carried out by the Operations Evaluation Department during 1999. Regional consultations are being planned for the second half of 1999 and will be complemented by country studies. A draft OED report is to be discussed by an 'Advisory Group' in the summer of 1999. The final report is scheduled to go to the World Bank's Board of Executive Directors later in the year. Jointly with other NGOs, the WRM submitted comments on the OED process 'design paper' to the Bank in early 1999.

Simultaneously, Bank staff from the Environment and Rural Development Departments will undertake a Stategy Development (SD) process, which will try to ascertain what the Bank's policy should be for the coming millennium. The results of the OED review are meant to feed into the SD which is meant to come to conclusions by April 2000. Details of the process can be found on the FPIRS website (see below for details). The Bank is contracting a number of agencies and individuals to write 15 'Analytic Papers' which are to be used to help stimulate discussions in the regional consultations and in the discussions of the draft Strategy paper. These are encouragingly broad and should stimulate a broad debate about the future of Bank involvement in forest-related policy and lending.

Participation: The Bank had initially contracted the IUCN to mediate the Bank's interactions with NGOs. However, NGOs including the WRM, objected strongly to this. As a result, the Bank's own NGO liaison unit will now help with NGO involvement but the IUCN is still being retained 'to support the Bank in developing and implementing a review process that remains transparent and open to all interested parties', though this has not until now resulted in positive feedback to and from NGOs.

Related Processes

Meanwhile the Bank is involved in two other related processes as a result of Wolfensohn's commitment to do something about forests. It has struck up an Alliance with the WorldWide Fund for Nature to promote the certification of logging and it has adopted WWF targets for setting up protected areas to cover 10% of the major forest types in all borrower countries.

This involves close collaboration with the WWF in forest policy discussions with WWF staff seconded to the Bank for liaison purposes.

The Bank is also engaged in a series of meetings with the CEOs of major forest industry companies to discuss best practice and see how industry can move towards this. Some NGOs and the IUCN have been involved in this process, but the early meetings conspicuously avoided addressing the social and political concerns. The WRM has written to these NGOs and the Bank to highlight the need to address social issues in discussions with industry and we have made recommendations that social NGOs, Indigenous Peoples Organisations and Trades Unions should be involved in future meetings.

Strategy Development Process

The Strategy Development process is being carried out simultaneously to the OED study when logically the review should precede it. There is a risk that Bank views will become entrenched and justified in the strategy development before the OED study can feed into it.

The degree to which the strategy development process will be open and participatory is not clear. Debate on early drafts and even later drafts does not seem to feature. In this sense the process seems less open than in 1994.

Plantations

Although tree plantations were not a major issue in the 1991 policy they had assumed far greater significance by 1994. The final review advocated much more lending for tree plantations under the guise of 'Resource Expansion'. The Bank's enthusiasm for CDM and joint implementation, related to the Kyoto Protocol, means that there is likely to be heavy emphasis on plantations in the strategy development even when there is growing evidence that large scale plantations result in highly negative social and environmental impacts, which can include forest clearing.

Recent Bank structural adjustment lending in South East Asia has also been aimed at expanding oil palm plantations. Tree crop estates are increasingly being developed by the private sector, creating resource conflicts with local communities and leading to increasing deforestation.

Forestry Projects

Bank staff are keen to lift the current prohibition on the Bank funding logging in primary moist tropical forests, whereas many NGOs want to see this prohibition extended to other forest types especially the remaining fragments of old growth, tropical dry, boreal and temperate forests.

Forestry is treated as a strategic sector of importance to climate, biodiversity and national development objectives. Local, social and poverty alleviation aspects take second place in these discussions.

Community Forestry and Collaborative Protected Area Management

While the Bank seems interested in funding community forestry and collaborative protected area projects, the Bank does not seem to emphasise the importance of tenure reforms, the special concerns of women and indigenous peoples.

The Bank may use the argument that it should fund community-based forest management as a justification for lifting the ban on funding logging in primary moist tropical forests.

Cross-Sectoral Issues

The most worrying issue remains the forestry focus of the Bank's approach. It seems that the World Bank's Vice Presidency for Environmentally and Socially Sustainable Development can't cope with the idea of having to reform the other sectors of the World Bank, although this is clearly necessary to address cross-sectoral causes of forest degradation and deforestation.

Recommendations to Influence the Process:

- Inform your country's Executive Directors as soon as possible of your concerns about the FPIRS
- Demand that the OED carried out participatory reviews of specific projects with affected communities to ensure that a real assessment of implementation is achieved
- Ensure that the FPIR looks at all sectors of Bank lending that affect forests and reviews the performance of all members of the Bank group including MIGA and the IFC
- Insist that the Strategy Development process be re-scheduled to follow after the completion of the FPIR
- Query the modalities for NGO and CBO participation in the process and demand clarity on the procedures that will be used to identify participants for regional and other consultations.

Recommendations to Influence the Bank's Policy:

- Create internal institutional mechanisms and oversight (p.e. incentives and penalties) to ensure World Bank staff compliance with the Forest Policy in all lending acitivities (projects, sectoral and structural adjustment programmes) as well as non-lending activities (Country-Assistance-Strategies, Technical Assistance Programmes).
- Develop the Bank's capacity for informed and effective participation by NGO and community based organisations in forest related projects and policy debates
- Address underlying causes of deforestation by addressing questions of indigenous rights, land security, agricultural stabilization, alternative income generating activities, and the impacts of other sectors including especially the energy, transport and mining sectors.
- Expand the ban on direct financing of logging in moist primary tropical forests to dry tropical forests, temperate and boreal forests.

- Ban funding of projects that damage or destroy any of the world's remaining primary forests, including logging operations, roads, mines, pipelines, tree plantations, etc..
- Suspend all loans to large-scale tree plantations until a specific and participatory assessment of the impacts of such plantations on forests, people and the environment is carried out. All types of tree plantations must be subject to this type of assessment, including plantations for pulpwood, timber, palm oil and carbon sinks.
- Establish a permanent mechanism to monitor and evaluate the World Bank's forest-related activities.

CROSS SECTORAL COMMITMENTS

- 'The Bank will give closer attention to infra-structural and other land-using projects to minimize their potentially negative effects on forests.'
- 'the Bank's economic and sector work should address forestry issues within a multisectoral context...(including) reforms of concession policies, forest revenue systems, fiscal, tax, and agricultural policies, infrastructural and land use planning procedures, land tenure systems and enforcement mechanisms'.
- 'forest sector policies advocated during project preparation will grow out of those put forward in the country economic memorandum, the national Environmental Action Plan, and the Consultative Group process... and as structural or sectoral adjustment inputs'. The guidelines advocate paying special attention to macroeconomic and policy issues such as revenue sharing, resource pricing, public investments in other sectors, institutional budgets, tarriffs, energy policy, incentive structures relating to land use.

BIBLIOGRAPHY

Bahati, J.: (1995). *Impact of Arboricidal Treatments on the Natural Regeneration of Species Composition in Budongo Forest*, Uganda. M. Sc., Makerere University.

Bethell, E.: (1998). *Vigilance in Foraging Chimpanzees*. University College, London University.

Fairgrieve, C.: (1995). *The Comparative Ecology of Blue Monkeys (Cercopithecus mitis stuhlmanni) in Logged and Unlogged Forest, Budongo Forest Reserve*, Uganda, Edinburgh University.

Hills, C.: (1998). *An Analysis of Phylogeny and Gene Flow in a Population of Pan troglodytes schweinfuthii*. M. Sc., Oxford University.

Johnson, K.: (1993). *Local Uses of Budongo's Forest Products*. M. Sc., Oxford University.

Johnson, K.: (1996). *Hunting in the Budongo Forest*, Uganda. SWARA.

Johnson, K.: (1996). *Local attitudes towards the Budongo Forest*, Western Uganda. Indigenous Knowledge & Development Monitor.

Miller, P. L.: (1993). *Some dragonflies of the Budongo Forest*, Western Uganda. Opusc.

Mork, K. & Solvang, R.: (1996). *Edge Effects on an Understorey Bird Community in Budongo Forest Reserve*, Uganda, Agricultural University of Norway.

Musamali, P.: (1996). *The Ecology, Diversity, and Relative Abundance of Small Mammals in Primary and Disturbed Forest Compartments in the Budongo Forest Reserve*. M. Sc., Makerere University.

Owiunji, I.: (1996). *The Long Term Effects of Forest Management on the Bird Community of Budongo Forest Reserve*, Uganda. M. Sc., Makerere University.

Reynolds, V.: (1965). *Budongo: A Forest and its Chimpanzees*, London, Methuen.

Reynolds, V. & Reynolds, F.: (1965). *Chimpanzees of the Budongo Forest*, New York, Rinehart and Winston.

INDEX

❑❑❑